IEE WIRING REGULATIONS

REGULATIONS FOR ELECTRICAL INSTALLATIONS

Fifteenth Edition 1981

Copies may be obtained from The Institution at the following address:

The Institution of Electrical Engineers,
P.O. Box No. 8, Hitchin, Herts. SG5 1RS.

Published by: The Institution of Electrical Engineers, London

British Library Cataloguing in Publication Data

Regulations for electrical installations.
15th ed.
1. Electric wiring, Interior
I. Title II. Institution of Electrical Engineers
621.319'24 TK327

ISBN 0-852962-35-5

Printed in the United Kingdom by A. McLay & Co. Ltd., Cardiff

Contents

		Page
Editions		iv
Wiring Regulations Committee, Constitution		v
Preface		vi
Notes on the plan of the 15th Edition		vii
PART 1 – SCOPE, OBJECT AND FUNDAMENTAL REQUIREMENTS FOR SAFETY		1
Chapter 11	Scope	2
Chapter 12	Object and effects	3
Chapter 13	Fundamental requirements for safety	4
PART 2 – DEFINITIONS		7
PART 3 – ASSESSMENT OF GENERAL CHARACTERISTICS		13
Chapter 31	Purposes, supplies and structure	14
Chapter 32	External influences	16
Chapter 33	Compatibility	16
Chapter 34	Maintainability	17
PART 4 – PROTECTION FOR SAFETY		18
Chapter 41	Protection against electric shock	19
Chapter 42	Protection against thermal effects	32
Chapter 43	Protection against overcurrent	34
Chapter 44	*(Reserved for future use)*	–
Chapter 45	*(Reserved for future use)*	–
Chapter 46	Isolation and switching	38
Chapter 47	Application of protective measures for safety	40
PART 5 – SELECTION AND ERECTION OF EQUIPMENT		52
Chapter 51	Common rules	54
Chapter 52	Cables, conductors, and wiring materials	59
Chapter 53	Switchgear	74
Chapter 54	Earthing arrangements and protective conductors	79
Chapter 55	Other equipment	88
PART 6 – INSPECTION AND TESTING		96
Chapter 61	Initial inspection and testing	97
Chapter 62	Alterations to installations	100
Chapter 63	Periodic inspection and testing	100
APPENDICES		101
1	Publications of the British Standards Institution to which reference is made in these Regulations	102
2	Statutory Regulations and associated memoranda	105
3	Explanatory notes on types of system earthing	107
4	Maximum demand and diversity	111
5	Standard circuit arrangements	113
6	Classification of external influences	117
7	An alternative method of compliance with Regulation 413-3 for circuits supplying socket outlets	124
8	Limitation of earth fault loop impedance for compliance with Regulation 543-1	127
9	Current-carrying capacities and associated voltage drops for cables and flexible cords	141
10	Notes on the selection of types of cable and flexible cord for particular uses and external influences	173
11	Notes on methods of support for cables, conductors and wiring systems	176
12	Cable capacities of conduit and trunking	180
13	Example of earthing arrangements and protective conductors	184
14	Check list for initial inspection of installations	185
15	Standard methods of testing	186
16	Forms of completion and inspection certificates	189
INDEX		193

Editions

The following editions have been published

FIRST EDITION	entitled 'Rules and Regulations for the Prevention of Fire Risks arising from Electric Lighting'. Issued in 1882.
SECOND EDITION	Issued in 1888.
THIRD EDITION	entitled 'General Rules recommended for Wiring for the Supply of Electrical Energy'. Issued in 1897.
FOURTH EDITION	Issued in 1903.
FIFTH EDITION	entitled 'Wiring Rules'. Issued in 1907.
SIXTH EDITION	Issued in 1911.
SEVENTH EDITION	Issued in 1916.
EIGHTH EDITION	entitled 'Regulations for the Electrical Equipment of Buildings'. Issued in 1924.
NINTH EDITION	Issued in 1927.
TENTH EDITION	Issued in 1934.
ELEVENTH EDITION	Issued in 1939.
ELEVENTH EDITION (REVISED)	Issued in 1943.
ELEVENTH EDITION (REVISED 1943)	Reprinted with minor Amendments, 1945.
	Supplement issued, 1946.
	Revised Section 8 issued, 1948.
TWELFTH EDITION	Issued in 1950.
	Supplement issued, 1954.
THIRTEENTH EDITION	Issued in 1955.
	Reprinted 1958, 1961, 1962 and 1964.
FOURTEENTH EDITION	Issued in 1966.
	Reprinted incorporating Amendments, 1968.
	Reprinted incorporating Amendments, 1969.
	Supplement on use in metric terms issued, 1969.
	Amendments issued, 1970.
	Reprinted in metric units incorporating Amendments, 1970.
	Reprinted 1972.
	Reprinted 1973.
	Amendments issued, 1974.
	Reprinted incorporating Amendments, 1974.
	Amendments issued, 1976.
	Reprinted incorporating Amendments, 1976.
FIFTEENTH EDITION	entitled 'Regulations for Electrical Installations'. Issued in 1981.

Wiring Regulations Committee

CONSTITUTION

as at September 1980

The President (ex officio)

A H Young DFH (Chairman)

D W M Latimer MA (Vice-Chairman)

P Bingley
G A Bowie DFH
P C Hoare
P M Hollingsworth M.Eng
E E Hutchings BSc
E J R Kay BSc
E J Sutton
J F Wilson
J R Worsley

and the Chairmen of the Power Division Professional Groups P4 – Industrial applications and processes, and P5 – Non-industrial utilisation and electrical building services (ex officio)

and	*nominated by*
W R Carmichael	Associated Offices Technical Committee
T B Rolls BA	Association of Consulting Engineers
R E Fenney	Association of Manufacturers of Domestic Electrical Appliances
E J Mathieson DFH	Association of Supervisory and Executive Engineers
E Coleman J Feenan	British Electrical and Allied Manufacturers' Association
J Vevers OBE	British Electrotechnical Approvals Board
M A E Butler MA	British Radio Equipment Manufacturers' Association
N Howard	British Railways Board
C E Mountcastle	British Standards Institution
J Wray	British Transport Docks Board
(Appointment pending)	Building Services Construction Forum
J H Gura	Chartered Institute of Building Services
M A Stothers	City and Guilds of London Institute
K H Caddick	Council of British Manufacturers and Contractors serving the Petroleum and Process Industries
P E Donnachie BSc	Department of the Environment
A T Baldock	Department of Energy
A Waldegrave	Department of Trade
N S Bryant BSc(Eng) E W G Bungay BSc	Electric Cable Makers' Confederation
J P Cutting	Electrical Contractors' Association
*F G Bennett	Electrical Contractors' Association of Scotland
D Rogers	Electrical, Electronic, Telecommunication and Plumbing Union
R G Parr BSc (Eng)	Electrical Research Association
D W Cowe BSc D V Ford G D Atkinson	Electricity Supply Industry in England and Wales
T C Jackson	Electricity Supply Industry in Northern Ireland
M C Whitfield	Electricity Supply Industry in Scotland
F H Familton	Engineer Surveyors' Association
A G Howell	Engineering Equipment Users' Association
G W A Nicholls	Fire Offices Committee
A Hall	Health & Safety Executive
P Laughton	Institution of Electrical and Electronics Technician Engineers
I F Davies BSc	Lighting Industry Federation
R D Tomlin	London Transport Executive
Major M C Travers RE	Ministry of Defence (Army)
R Hartill BSc	National Coal Board
C D Kinloch DFH	National Inspection Council for Electrical Installation Contracting
R V Hogan	Oil Companies Materials Association
J O Colyer, BSc(Eng)	Post Office
(Appointment pending)	Royal Institute of British Architects
R A E Quartermaine	Scottish Development Department

*Alternate: T E Marshall

Preface

This edition was issued on 30th March 1981. It is intended to supersede the Fourteenth Edition on 1 January 1983, but until that date both editions are intended to have equal validity.

It is arranged according to the new plan for Publication 364 – 'Electrical installations of buildings' of the International Electrotechnical Commission (IEC), so far as concerns the general arrangement of the main parts, chapters, and sections but the clause numbering does not correspond to that of the international publication.

The IEC plan has also been adopted for the purposes of the corresponding work of the European Committee for Electrotechnical Standardization (CENELEC) for the harmonization of the national wiring regulations of the member countries of the European Economic Community and the European Free Trade Association.

The IEE Wiring Regulations Committee, acting on behalf of, and through, the British Electrotechnical Committee, provide the British contribution to the work for IEC Publication 364 and the corresponding CENELEC work.

In this edition, so far as is practicable, account has been taken of the technical substance of the parts of IEC Publication 364 so far published*, and of the corresponding agreements reached in CENELEC. In particular, this edition takes account of the following CENELEC Harmonization Documents which have been accepted by the British Electrotechnical Committee on the advice of the Wiring Regulations Committee:

H.D. 193	–	Voltage Bands
H.D. 308	–	Identification and use of cores of flexible cables
H.D. 384.1	–	Electrical installations of buildings – Scope
H.D. 384.4.41	–	Electrical installations of buildings – Protection against electric shock
H.D. 384.4.43	–	Electrical installations of buildings – Protection against overcurrent
H.D. 384.4.473	–	Electrical installations of buildings – Application of measures for protection against overcurrent.

As the IEC and CENELEC work is at a relatively early stage, certain parts of this edition are essentially a rearrangement and factual updating of the content of the previous (14th, as amended to April 1976) edition, where no corresponding international results are yet available.

The Regulations will be amended from time to time to take account of further progress of the international work and other developments, the arrangement of parts, chapters, and sections being intended to facilitate this. The publication of a further edition will be considered when IEC Publication 364 and the corresponding CENELEC work are nearer completion.

The opportunity has also been taken to revise certain regulations for greater clarity or to take account of technical developments.

Considerable reference is made throughout these Regulations to publications of the British Standards Institution, both specifications and codes of practice. Appendix 1 lists these publications and gives their full titles whereas throughout these Regulations they are referred to only by their numbers.

Where reference is made to a British Standard in these Regulations, and the British Standard concerned takes account of a CENELEC Harmonization Document, it is understood that the reference is to be read as relating also to any foreign standard similarly based on that Harmonization Document, provided it is verified that any differences between the two standards would not result in a lesser degree of safety than that achieved by compliance with the British Standard.

A similar understanding is applicable to national standards based on IEC standards but as national deviations are not required to be listed in such standards, special care should be exercised in assessing any national differences.

*Details may be obtained from the Secretary of the Institution.

Notes on the plan of the 15th edition

This edition is based on the plan agreed internationally for the arrangement of safety rules for electrical installations.

In the numbering system used, the first digit signifies a Part, the second digit a Chapter, the third digit a Section, and the subsequent digits the regulation number. For example, the Section number 413 is made up as follows:

PART 4 – PROTECTION FOR SAFETY

Chapter 41 (first chapter of Part 4) – Protection against electric shock
Section 413 (third section of Chapter 41) – Protection against indirect contact.

Part 1 sets out fundamental requirements for safety that are applicable to all installations.

Part 2 defines the sense in which certain terms are used throughout the Regulations.

The subjects of the subsequent parts are as indicated below:

Part No.	Subject
3	Identification of the characteristics of the installation that will need to be taken into account in choosing and applying the requirements of the subsequent Parts. These characteristics may vary from one part of an installation to another, and should be assessed for each location to be served by the installation.
4	Description of the basic measures that are available, for the protection of persons, livestock, and property against the hazards that may arise from the use of electricity. Chapters 41 to 46 each deal with a particular hazard. Chapter 47 deals in more detail with, and qualifies, the practical application of the basic protective measures, and is divided into Sections whose numbering corresponds to the numbering of the preceding chapters; thus Section 471 needs to be read in conjunction with Chapter 41, Section 473 with Chapter 43, and so on.
5	Precautions to be taken in the selection and erection of the equipment of the installation. Chapter 51 relates to equipment generally and Chapters 52 to 55 to particular types of equipment.
6	Inspection and testing.

The sequence of the plan should be followed in considering the application of any particular requirement of these Regulations. The general index provides a ready reference to particular regulations by subject, but in applying any one regulation the requirements of related regulations should be borne in mind. Cross-references are provided, and the index is arranged, to facilitate this.

PART 1

Scope, object and fundamental requirements for safety

CONTENTS

CHAPTER 11 – SCOPE

11–1	General
11–2	Voltage ranges
11–3	Exclusions from scope
11–4	Equipment
11–5	Temporary prefabricated installations

CHAPTER 12 – OBJECT AND EFFECTS

12–1 and 12–2	General
12–3	Relationship with Statutory Regulations
12–4	Use of established materials, equipment and methods
12–5	New materials, inventions and designs
12–6	Assessment of New Techniques (ANT) Scheme
12–7	Installations in premises subject to licensing
12–8	Status of notes to the Regulations
12–9	Date of validity

CHAPTER 13 – FUNDAMENTAL REQUIREMENTS FOR SAFETY

13–1	Workmanship and materials
13–2 to 13–6	General
13–7	Overcurrent protective devices
13–8 to 13–11	Precautions against earth leakage and earth fault currents
13–12 and 13–13	Position of protective devices and switches
13–14 and 13–15	Isolation and switching
13–16	Accessibility of equipment
13–17 and 13–18	Precautions in adverse conditions
13–19	Additions and alterations to an installation
13–20	Inspection and testing

PART 1

Scope, object and fundamental requirements for safety

CHAPTER 11

SCOPE

General

11–1 These Regulations relate principally to the design, selection, erection, inspection and testing of electrical installations, whether permanent or temporary, in and about buildings generally.

They also relate to electrical installations of:

(i) agricultural and horticultural premises,
(ii) construction sites,
(iii) caravans and their sites.

It is recognised that these Regulations are commonly used for other purposes and it is not intended to discourage their application to other types of installation for which they may be mainly suitable; in that case, however, these Regulations may need to be modified or supplemented.

NOTE – For further information concerning construction sites, see CP1017.

Voltage ranges

11–2 Installations utilising the following nominal voltage ranges are dealt with:

(i) Extra-low voltage – Normally not exceeding 50V a.c. or 120V d.c. whether between conductors or to Earth.

(ii) low voltage – Normally exceeding extra-low voltage but not exceeding 1000V a.c. or 1500V d.c. between conductors or 600V a.c. or 900V d.c. between any conductor and Earth.

Regulations are also included for certain installations operating at voltages exceeding low voltage, for discharge lighting and electrode boilers.

Exclusions from scope

11–3 These Regulations do not apply to:

(i) systems for distribution of energy to the public, or to power generation and transmission for such systems.

NOTE – In Great Britain, such systems are subject to the Electricity Supply Regulations, 1937. In Northern Ireland, corresponding regulations by the Secretary of State for Northern Ireland apply. Outside the United Kingdom, reference should be made to any corresponding requirements.

(ii) those aspects of installations in potentially explosive atmospheres relating to methods of dealing with the explosion hazard which are specified in BS 5345 and CP 1003, or in premises where the fire risks are of an unusual character so as to require special measures.

(iii) those parts of telecommunications (e.g. radio, telephone, bell, call and sound distribution and data transmission), fire alarm, intruder alarm and emergency lighting circuits and equipment that are fed from a safety source complying with Regulation 411–3. Requirements for the segregation of other circuits from such circuits are, however, included.

NOTE – For fire alarm systems see BS 5839 and for emergency lighting of premises see BS 5266.

(iv) electric traction equipment.

(v) electrical equipment of motor vehicles, except those to which the requirements of these Regulations concerning caravans are applicable.

(vi) electrical equipment on board ships.

NOTE – See the 'Regulations and Recommendations for the Electrical and Electronic Equipment of Ships'.*

*IEE Publications Department.

(vii) electrical equipment on offshore installations.

NOTE – IEE Regulations for offshore installations are in course of preparation.

(viii) electrical equipment of aircraft.

NOTE – See the British Civil Airworthiness Requirements.+

(ix) installations at mines and quarries.

(x) radio interference suppression equipment, except so far as it affects safety of the electrical installation.

(xi) lightning protection of buildings.

NOTE – For guidance on protection of buildings against lightning see CP 326.

Equipment

11–4 These Regulations apply to items of electrical equipment only so far as selection and application of the equipment in the installations are concerned. These Regulations do not deal with requirements for the construction of prefabricated assemblies of electrical equipment, where these assemblies comply with appropriate specifications.

Temporary prefabricated installations

11–5 For prefabricated installations for temporary use and frequent erection, only those Regulations concerning design and selection apply.

NOTE – For fairgrounds see the Home Office Guide to Safety at Fairs.†

+Civil Aviation Authority
†H.M. Stationery Office.

CHAPTER 12

OBJECT AND EFFECTS

General

12–1 These Regulations are designed to provide safety, especially from fire, shock and burns.

12–2 These Regulations are intended to be cited in their entirety if referred to in any contract. They are not intended to take the place of a detailed specification or to instruct untrained persons or to provide for every circumstance. Installations of a difficult or special character will require the advice of a suitably qualified electrical engineer.

Relationship with Statutory Regulations

12–3 Compliance with Chapter 13 of these Regulations will, in general, satisfy the requirements of the statutory regulations listed in Appendix 2, where the latter specifically apply to the provisions for electrical installations in buildings.

Parts 3 to 6 of these Regulations set out in greater detail methods and practices which are regarded as meeting the requirements of Chapter 13. Any departure from those Parts needs to be the subject of special consideration and shall be noted in the completion certificate specified in Part 6.

Use of established materials, equipment and methods

12–4 Only established materials, equipment and methods are considered, but it is not intended to discourage invention or to exclude other materials, equipment and methods affording an equivalent degree of safety which may be developed in the future.

New materials, inventions and designs

12–5 Where the use of a material, invention or design leads to departures from these Regulations, the resulting degree of safety of the installation shall be not less than that obtained by compliance with these Regulations.

Such departures shall be the subject of a written specification of a competent body, or competent person or persons, and the installation shall not be described as complying with these Regulations.

Assessment of New Techniques (ANT) Scheme

12–6 Where a new technique not envisaged in these Regulations is to be offered for general use, and there is a need for a more formalised assessment that the technique is capable of affording a degree of safety not less than that achieved by compliance with Parts 3 to 6 of these Regulations, such cases are suitable for consideration under the Assessment of New Techniques (ANT) Scheme operated by the Wiring Regulations Committee. An installation incorporating the subject of an ANTS certificate shall not be described as complying with these Regulations.

NOTE – Details of the ANT Scheme and a list of current ANTS Certificates may be obtained from the Secretary of the Institution.

Installations in premises subject to licensing

12–7 For installations in premises over which a licensing or other authority exercises a statutory control, the requirements of that authority shall be ascertained.

NOTE – For further information on statutory regulations see Appendix 2.

Status of notes to the Regulations

12–8 Notes to these Regulations do not form part of the Regulations but may indicate ways in which the requirements can be met.

Date of validity

12–9 This edition supersedes the Fourteenth Edition on 1 January 1983. Until that date both editions are intended to have equal validity.

CHAPTER 13

FUNDAMENTAL REQUIREMENTS FOR SAFETY

Workmanship and materials

13–1 Good workmanship and proper materials shall be used.

General

13–2 All equipment shall be constructed, installed and protected, and shall be capable of being maintained, inspected and tested, so as to prevent danger so far as is reasonably practicable.

13–3 All equipment shall be suitable for the maximum power demanded by the current-using equipment when it is functioning in its intended manner.

13–4 All electrical conductors shall be of sufficient size and current-carrying capacity for the purposes for which they are intended.

13–5 All conductors shall either–

(i) be so insulated, and where necessary further effectively protected, or

(ii) be so placed and safeguarded,

as to prevent danger, so far as is reasonably practicable.

13–6 Every electrical joint and connection shall be of proper construction as regards conductance, insulation, and mechanical strength and protection.

Overcurrent protective devices

13–7 Where necessary to prevent danger, every installation and every circuit thereof shall be protected against overcurrent by devices which–

(i) will operate automatically at values of current which are suitably related to the safe current ratings of the circuit, and

(ii) are of adequate breaking capacity and, where appropriate, making capacity, and

(iii) are suitably located and are constructed so as to prevent danger from overheating, arcing or the scattering of hot particles when they come into operation and to permit ready restoration of the supply without danger.

NOTE – Where the supply undertaking provides switchgear or fusegear at the origin of the installation it may not be necessary to duplicate the means of overcurrent protection for that part of the installation between its origin and the main distribution point of the installation where the next step for overcurrent protection is provided.

Precautions against earth leakage and earth fault currents

13–8 Where metalwork of electrical equipment, other than current-carrying conductors, may become charged with electricity in such a manner as to cause danger if the insulation of a conductor should become defective or if a fault should occur in any equipment–

(i) the metalwork shall be earthed in such a manner as will cause discharge of electrical energy without danger, or

(ii) other equally effective precautions shall be taken to prevent danger.

13–9 Every circuit shall be arranged so as to prevent the persistence of dangerous earth leakage currents.

NOTE – In Great Britain see also Regulation 26 of the Electricity Supply Regulations 1937.

13–10 Where metalwork is earthed in accordance with Regulation 13-8(i) the circuits concerned shall be protected against the persistence of dangerous earth fault currents by–

(i) the overcurrent protective devices required by Regulation 13-7, or

(ii) a residual current or voltage operated device or equally effective device.

The method described in item (ii) above shall be used whenever the prospective earth fault current is insufficient to cause prompt operation of the overcurrent protective devices.

13–11 Where necessary to prevent danger, where metalwork of electrical equipment is earthed for compliance with Regulation 13-8(i) and is accessible simultaneously with substantial exposed metal parts of other services, the latter parts shall be effectively connected to the main earthing terminal of the installation.

Position of protective devices and switches

13–12 No fuse, or circuit breaker other than a linked circuit breaker, shall be inserted in an earthed neutral conductor, and any linked circuit breaker inserted in an earthed neutral conductor shall be arranged to break also all the related phase conductors.

13–13 Every single-pole switch shall be inserted in the phase conductor only, and any switch connected in an earthed neutral conductor shall be a linked switch and shall be arranged to break also all the related phase conductors.

NOTE – With regard to Regulations 13–12 and 13–13, for a supply given in accordance with the Electricity Supply Regulations, 1937, it may be assumed that the connection with Earth of the neutral of the supply is permanent. Outside Great Britain, confirmation should be sought from the supply undertaking that the supply conforms to requirements corresponding to those of the Electricity Supply Regulations, 1937, in this respect.

Isolation and switching

13–14 Effective means, suitably placed for ready operation, shall be provided so that all voltage may be cut off from every installation, from every circuit thereof and from all equipment, as may be necessary to prevent or remove danger.

NOTE – Where the supply undertaking provides switchgear or fusegear at the origin of the installation it may not be necessary to duplicate the means of isolation for that part of the installation between its origin and the main distribution point of the installation where the next step for isolation is provided.

13–15 For every electric motor an efficient means of disconnection shall be provided which shall be readily accessible, easily operated and so placed as to prevent danger.

Accessibility of equipment

13–16 Every piece of equipment which requires operation or attention by a person in normal use shall be so installed that adequate and safe means of access and working space are afforded for such operation or attention.

Precautions in adverse conditions

13–17 All equipment likely to be exposed to weather, corrosive atmospheres, or other adverse conditions, shall be so constructed or protected as may be necessary to prevent danger arising from such exposure.

13–18 All equipment in surroundings susceptible to risk of fire or explosion shall be so constructed or protected, and such other special precautions shall be taken, as may be necessary to prevent danger.

NOTE– See Appendix 2 for relevant statutory regulations, see also BS 5345 and CP 1003.

Additions and alterations to an installation

13–19 No additions or alterations, temporary or permanent, shall be made to an existing installation, unless it has been ascertained that the ratings and the condition of any existing equipment (including that of the supply undertaking) which will have to carry any additional load is adequate for the altered circumstances and that the earthing arrangements are also adequate.

Inspection and testing

13–20 On completion of an installation or an extension or alteration of an installation, appropriate tests and inspection shall be made, to verify so far as reasonably practicable that the requirements of Regulations 13-1 to 13-19 have been met.

PART 2

Definitions

NOTE – The following definitions indicate the sense in which the terms defined are used in these Regulations. Some of these definitions are aligned with those given in BS 4727 – 'Glossary of electrotechnical, power, telecommunication, electronics, lighting and colour terms'. Other terms not defined herein are used in the sense defined in BS 4727.

Accessory. A device, other than current-using equipment, associated with such equipment or with the wiring of an installation.

Ambient temperature. The temperature of the air or other medium where the equipment is to be used.

Appliance. An item of current-using equipment other than a luminaire or an independent motor.

Arm's reach. A zone of accessibility to touch, extending from any point on a surface where persons usually stand or move about, to the limits which a person can reach with his hand in any direction without assistance.

Barrier. A part providing a defined degree of protection against contact with live parts, from any usual direction of access.

Basic insulation. Insulation applied to live parts to provide basic protection against electric shock.

NOTE – Basic insulation does not necessarily include insulation used exclusively for functional purposes.

Bonding conductor. A protective conductor providing equipotential bonding.

Bunched. Cables are said to be bunched when two or more are contained within a single conduit, duct, ducting, or trunking or, if not enclosed, are not separated from each other.

Cable coupler. A means enabling the connection, at will, of two flexible cables. It consists of a connector and a plug.

Cable ducting. A manufactured enclosure of metal or insulating material, other than conduit or cable trunking, intended for the protection of cables which are drawn-in after erection of the ducting, but which is not specifically intended to form part of a building structure.

Caravan. Any structure designed or adapted for human habitation which is capable of being moved from one place to another (whether by being towed or being transported on a motor vehicle or trailer) and any other motor vehicle so designed or adapted but not including–

(a) any railway rolling stock which is for the time being on rails forming part of a railway system, or

(b) any tent.

NOTE – For multi-unit structures see the amendment to the definition for caravans in the Caravan Sites Act 1968.*

Caravan site. Land on which a caravan is stationed for the purposes of human habitation and land which is used in conjunction with land on which a caravan is so stationed.

Cartridge fuse link. A device comprising a fuse element or several fuse elements connected in parallel enclosed in a cartridge usually filled with an arc-extinguishing medium and connected to terminations. The fuse link is the part of a fuse which requires replacing after the fuse has operated.

Circuit. An assembly of electrical equipment supplied from the same origin and protected against overcurrent by the same protective device(s). For the purposes of Chapter 52 of these Regulations, certain types of circuit are categorised as follows:

Category 1 circuit – A circuit (other than a fire alarm or emergency lighting circuit) operating at low voltage and supplied directly from a mains supply system.

Category 2 circuit – With the exception of fire alarm and emergency lighting circuits, any circuit for telecommunication (e.g. radio, telephone, sound distribution, intruder alarm, bell and call, and

*H.M. Stationery Office

data transmission circuits) which is supplied from a safety source complying with Regulation 411-3.

Category 3 circuit – A fire alarm circuit or an emergency lighting circuit.

Circuit breaker. A mechanical switching device capable of making, carrying and breaking currents under normal circuit conditions and also of making, carrying for a specified time, and breaking currents under specified abnormal circuit conditions such as those of short circuit.

NOTE – A circuit breaker is usually intended to operate infrequently, although some types are suitable for frequent operation.

Circuit protective conductor. A protective conductor connecting exposed conductive parts of equipment to the main earthing terminal.

Class I equipment. Equipment in which protection against electric shock does not rely on basic insulation only, but which includes means for the connection of exposed conductive parts to a protective conductor in the fixed wiring of the installation.

NOTE – For information on classification of equipment with regard to means provided for protection against electric shock see BS 2754.

Class II equipment. Equipment in which protection against electric shock does not rely on basic insulation only, but in which additional safety precautions such as supplementary insulation are provided, there being no provision for the connection of exposed metalwork of the equipment to a protective conductor, and no reliance upon precautions to be taken in the fixed wiring of the installation.

NOTE – For information on classification of equipment with regard to the means provided for protection against electric shock see BS 2754.

Confined conductive location. A location having surfaces which are mainly composed of extraneous conductive parts and which are of such dimensions that movement is restricted to such an extent that contact with surfaces is difficult to avoid (e.g. in a boiler).

Connector. The part of a cable coupler or of an appliance coupler which is provided with female contacts and is intended to be attached to the flexible cable connected to the supply.

Current-carrying capacity of a conductor. The maximum current which can be carried by a conductor under specified conditions without its steady state temperature exceeding a specified value.

Current-using equipment. Equipment which converts electrical energy into another form of energy, such as light, heat, or motive power.

Danger. Danger to health or danger to life or limb from shock, burn, or injury from mechanical movement to persons (and livestock where present), or from fire, attendant upon the use of electrical energy.

Design current (of a circuit). The magnitude of the current intended to be carried by the circuit in normal service.

Direct contact. Contact of persons or livestock with live parts which may result in electric shock.

Double insulation. Insulation comprising both basic insulation and supplementary insulation.

Duct. A closed passage way formed underground or in a structure and intended to receive one or more cables which may be drawn in.

Ducting. *(See Cable ducting).*

Earth. The conductive mass of the Earth, whose electric potential at any point is conventionally taken as zero.

Earth electrode. A conductor or group of conductors in intimate contact with, and providing an electrical connection to, Earth.

Earth electrode resistance. The resistance of an earth electrode to Earth.

Earth fault loop impedance. The impedance of the earth fault current loop (phase to earth loop) starting and ending at the point of earth fault.

NOTE – See Appendix 15 for a description of the constituent parts of an earth fault current loop.

Earth leakage current. A current which flows to Earth, or to extraneous conductive parts, in a circuit which is electrically sound.

NOTE – This current may have a capacitive component including that resulting from the deliberate use of capacitors.

Earthed concentric wiring. A wiring system in which one or more insulated conductors are completely surrounded throughout their length by a conductor, for example a sheath, which acts as a PEN conductor.

Earthing conductor. A protective conductor connecting a main earthing terminal of an installation to an earth electrode or to other means of earthing.

Electric shock. A dangerous pathophysiological effect resulting from the passing of an electric current through a human body or an animal.

Electrical equipment. (abbr: *Equipment*). Any item for such purposes as generation, conversion, transmission, distribution or utilisation of electrical energy, such as machines, transformers, apparatus, measuring instruments, protective devices, wiring materials, accessories, and appliances.

Electrical installation. (abbr: *Installation*). An assembly of associated electrical equipment to fulfil a specific purpose and having certain co-ordinated characteristics.

Electrically independent earth electrodes. Earth electrodes located at such a distance from one another that the maximum current likely to flow through one of them does not significantly affect the potential of the other(s).

Electrode boiler (or electrode water heater). Equipment for the electrical heating of water or electrolyte by the passage of an electric current between electrodes immersed in the water or electrolyte.

Emergency switching. Rapid cutting off of electrical energy to remove any hazard to persons, livestock, or property which may occur unexpectedly.

Enclosure. A part providing an appropriate degree of protection of equipment against certain external influences and a defined degree of protection against contact with live parts from any direction.

Equipment. (abbr: *see Electrical equipment*).

Equipotential bonding. Electrical connection putting various exposed conductive parts and extraneous conductive parts at a substantially equal potential.

Exposed conductive part. A conductive part of equipment which can be touched and which is not a live part but which may become live under fault conditions.

External influence. Any influence external to an electrical installation which affects the design and safe operation of that installation.

Extraneous conductive part. A conductive part liable to transmit a potential including earth potential and not forming part of the electrical installation.

Factory-built assembly (of LV switchgear and control gear). An assembly built and assembled under the responsibility of the manufacturer, and conforming to an established type or system, without deviations likely to significantly influence the performance, from the typical assembly verified to be in accordance with the relevant British Standard.

NOTE – For various reasons, e.g. transport or production, certain steps of assembly may be made in a place outside the factory of the manufacturer of the factory-built assembly. Such assemblies are considered as factory-built provided the assembly is performed in accordance with the manufacturer's instructions, in such a manner that compliance with the relevant British Standard is assured, including submission to applicable routine tests.

Final circuit. A circuit connected directly to current-using equipment, or to a socket outlet or socket outlets or other outlet points for the connection of such equipment.

Fixed equipment. Equipment fastened to a support or otherwise secured in a specific location.

Functional earthing. Connection to earth necessary for proper functioning of electrical equipment.

Fuse element. A part of a fuse designed to melt when the fuse operates.

Fuse link. A part of a fuse, including the fuse element(s), which requires replacement by a new or renewable fuse link after the fuse has operated and before the fuse is put back into service.

Indirect contact. Contact of persons or livestock with exposed conductive parts made live by a fault and which may result in electric shock.

Installation. (abbr: *see Electrical installation*).

Instructed person. A person adequately advised or supervised by skilled persons to enable him to avoid dangers which electricity may create.

Insulating floor (or wall). A floor (or wall) such that, in the event of direct contact with a live part, a person standing on the floor (or touching the wall) cannot be traversed by a shock current flowing to the floor (or wall).

Insulation. Suitable non-conductive material enclosing, surrounding, or supporting a conductor.

NOTE – See also the definitions for basic insulation, double insulation, reinforced insulation, and supplementary insulation.

Isolation. Cutting off an electrical installation, a circuit, or an item of equipment from every source of electrical energy.

Live part. A conductor or conductive part intended to be energised in normal use, including a neutral conductor but, by convention, not a PEN conductor.

Luminaire. Equipment which distributes, filters, or transforms the light from one or more lamps, and which includes any parts necessary for supporting, fixing and protecting the lamps, but not the lamps themselves, and, where necessary, circuit auxiliaries together with the means for connecting them to the supply.

NOTE – For the purposes of these Regulations a batten lampholder, or a lampholder suspended by a flexible cord, is a luminaire.

Main earthing terminal. The terminal or bar provided for the connection of protective conductors, including equipotential bonding conductors, and conductors for functional earthing if any, to the means of earthing.

Mechanical maintenance. The replacement, refurbishment or cleaning of lamps and non-electrical parts of equipment, plant and machinery.

Neutral conductor. A conductor connected to the neutral point of a system and contributing to the transmission of electrical energy.

NOTE – The term also means the equivalent conductor of a d.c. system unless otherwise specified in these Regulations.

Nominal voltage. *(See Voltage, nominal).*

Obstacle. A part preventing unintentional contact with live parts but not preventing deliberate contact.

Origin of an installation. The position at which electrical energy is delivered to an installation.

Overcurrent. A current exceeding the rated value. For conductors the rated value is the current-carrying capacity.

Overcurrent detection. A method of establishing that the value of current in a circuit exceeds a predetermined value for a specified length of time.

Overload current. An overcurrent occurring in a circuit which is electrically sound.

PEN conductor. A conductor combining the functions of both protective conductor and neutral conductor.

Phase conductor. A conductor of an a.c. system for the transmission of electrical energy, other than a neutral conductor.

NOTE – The term also means the equivalent conductor of a d.c. system unless otherwise specified in these Regulations.

Plug. A device, provided with contact pins, which is intended to be attached to a flexible cable, and which can be engaged with a socket outlet or with a connector.

Point (in wiring). A termination of the fixed wiring intended for the connection of current-using equipment.

Portable equipment. Equipment which is moved while in operation or which can easily be moved from one place to another while connected to the supply.

Protective conductor. A conductor used for some measures of protection against electric shock and intended for connecting together any of the following parts:

– exposed conductive parts,
– extraneous conductive parts,
– the main earthing terminal,
– earth electrode(s),
– the earthed point of the source, or an artificial neutral.

Reinforced insulation. Single insulation applied to live parts, which provides a degree of protection against electric shock equivalent to double insulation under the conditions specified in the relevant standard.

NOTE – The term 'single insulation' does not imply that the insulation must be one homogeneous piece. It may comprise several layers which cannot be tested singly as supplementary or basic insulation.

Residual current device. A mechanical switching device or association of devices intended to cause the opening of the contacts when the residual current attains a given value under specified conditions.

Residual operating current. Residual current which causes the residual current device to operate under specified conditions.

Resistance area (for an earth electrode only). The surface area of ground (around an earth electrode) on which a significant voltage gradient may exist.

Ring final circuit. A final circuit arranged in the form of a ring and connected to a single point of supply.

Shock current. A current passing through the body of a person or an animal and having characteristics likely to cause dangerous pathophysiological effects.

Short circuit current. An overcurrent resulting from a fault of negligible impedance between live conductors having a difference in potential under normal operating conditions.

Simultaneously accessible parts. Conductors or conductive parts which can be touched simultaneously by a person or, where applicable, by livestock.

NOTES – 1. Simultaneously accessible parts may be–
– live parts,
– exposed conductive parts,
– extraneous conductive parts,
– protective conductors,
– earth electrodes.

2. This term applies for livestock in locations specifically intended for these animals.

Skilled person. A person with technical knowledge or sufficient experience to enable him to avoid dangers which electricity may create.

Socket outlet. A device, provided with female contacts, which is intended to be installed with the fixed wiring, and intended to receive a plug.

NOTE – A luminaire track system complying with BS 4533 is not regarded as a socket outlet system.

Space factor. The ratio (expressed as a percentage) of the sum of the overall cross-sectional areas of cables (including insulation and any sheath) to the internal cross-sectional area of the conduit or other cable enclosure in which they are installed. The effective overall cross-sectional area of a non-circular cable is taken as that of a circle of diameter equal to the major axis of the cable.

Spur. A branch cable connected to a ring or radial final circuit.

Stationary equipment. Equipment which is either fixed, or equipment having a mass exceeding 18kg and not provided with a carrying handle.

Supplementary insulation. Independent insulation applied in addition to basic insulation in order to provide protection against electric shock in the event of a failure of basic insulation.

Switch. A mechanical switching device capable of making, carrying and breaking current under normal circuit conditions, which may include specified operating overload conditions, and also of carrying for a specified time currents under specified abnormal circuit conditions such as those of short circuit.

NOTE – A switch may also be capable of making, but not breaking, short circuit currents.

Switch, linked. A switch the contacts of which are so arranged as to make or break all poles simultaneously or in a definite sequence.

Switchboard. An assembly of switchgear with or without instruments, but the term does not apply to a group of local switches in a final circuit.

NOTE – In the Electricity (Factories Act) Special Regulations 1908 and 1944, the term 'Switchboard' includes a distribution board.

Switchgear. An assembly of main and auxiliary switching apparatus for operation, regulation, protection or other control of electrical installations.

System. An electrical system consisting of a single source of electrical energy and an installation. For certain purposes of these Regulations, types of system are identified as follows, depending upon the relationship of the source, and of exposed conductive parts of the installation, to Earth:

- **TN system** A system having one or more points of the source of energy directly earthed, the exposed conductive parts of the installation being connected to that point by protective conductors. Three types of TN systems are recognised as follows:
  - **TN-C system,** in which neutral and protective functions are combined in a single conductor throughout the system,
  - **TN-S system,** having separate neutral and protective conductors throughout the system,
  - **TN-C-S system,** in which neutral and protective functions are combined in a single conductor in part of the system.
- **TT system** A system having one point of the source of energy directly earthed, the exposed conductive parts of the installation being connected to earth electrodes electrically independent of the earth electrodes of the source.
- **IT system** A system having no direct connection between live parts and Earth, the exposed conductive parts of the electrical installation being earthed.

NOTE – See Appendix 3 for further explanation of the significance of the terms used to describe types of system.

Trunking (for cables). A system of enclosures for the protection of cables, normally of square or rectangular cross section, of which one side is removable or hinged.

Voltage, nominal. Voltage by which an installation (or part of an installation) is designated. The following ranges of nominal voltage (r.m.s. values for a.c.) are defined:

- **Extra low** Normally not exceeding 50V a.c. or 120V d.c., whether between conductors or to Earth.
- **Low** Normally exceeding extra-low voltage but not exceeding 1000V a.c. or 1500V d.c. between conductors, or 600V a.c. or 900V d.c. between conductors and Earth.

NOTE – The actual voltage of the installation may differ from the nominal value by a quantity within normal tolerances.

PART 3

Assessment of general characteristics

CONTENTS

PART 3

Assessment of general characteristics

General

300–1 An assessment shall be made of the following characteristics of the installation in accordance with the chapters indicated:

(i) the purpose for which the installation is intended to be used, its general structure, and its supplies (Chapter 31),

(ii) the external influences to which it is to be exposed (Chapter 32),

(iii) the compatibility of its equipment (Chapter 33),

(iv) its maintainability (Chapter 34).

These characteristics shall be taken into account in the choice of methods of protection for safety (see Part 4) and the selection and erection of equipment (see Part 5).

CHAPTER 31

PURPOSES, SUPPLIES AND STRUCTURE

311 MAXIMUM DEMAND AND DIVERSITY

Maximum demand

311–1 The maximum demand of the installation, expressed as a current value, shall be assessed.

Diversity

311–2 In determining the maximum demand of an installation or parts thereof, diversity may be taken into account.

NOTE to 311–1 and 311–2 – Appendix 4 gives some information on the determination of the maximum demand of an installation and includes the current demand to be assumed for commonly used equipment together with guidance on the application of allowances for diversity.

312 ARRANGEMENT OF LIVE CONDUCTORS AND TYPE OF EARTHING

General

312–1 The characteristics mentioned in Regulations 312–2 and 312–3 shall be ascertained in order to determine which methods of protection for safety, selected from Part 4 of these Regulations, will be appropriate.

Number and types of live conductors

312–2 The number and types of live conductors (e.g. single-phase two-wire a.c., three-phase four-wire a.c.) shall be assessed, both for the source of energy and for the circuits to be used within the installation. Where the source of energy is provided by a supply undertaking, that undertaking shall be consulted if necessary.

Types of earthing arrangement

312–3 The type of earthing arrangement or arrangements to be used for the installation shall be determined.

NOTES: 1 – See the definition for 'system'. See also Appendix 3 for explanatory notes on types of system earthing and on the distinction between 'source of energy' and 'installation'.

2 – The choice of the arrangements may be limited by the characteristics of the source of energy, in particular, any facilities for earthing.

313 NATURE OF SUPPLY

General

313–1 The following characteristics of the supply or supplies shall be ascertained for an external supply and shall be determined for a private source:

(i) nominal voltage(s),

(ii) nature of current and frequency,

(iii) prospective short circuit current at the origin of the installation,

(iv) type and rating of the overcurrent protective device acting at the origin of the installation,

(v) suitability for the requirements of the installation, including the maximum demand.

(vi) the earth loop impedance of that part of the system external to the installation.

NOTE – As regards item (vi) it may only be possible to ascertain an expected maximum value.

Supplies for safety services and standby purposes

313–2 Where a supply for safety services or standby purposes is required by the person specifying the installation, the characteristics of the source or sources of any such supply shall be assessed. These sources shall have adequate capacity and rating for the operation specified.

Where the normal source of energy is to be provided by a supply undertaking, the supply undertaking shall be consulted regarding switching arrangements for safety and standby supplies, especially where the various sources are intended to operate in parallel.

NOTE – For information concerning emergency lighting of premises see BS 5266, and for fire alarm systems see BS 5839.

314 INSTALLATION CIRCUIT ARRANGEMENTS

314–1 Every installation shall be divided into circuits as necessary to–

(i) avoid danger and minimise inconvenience in the event of a fault, and

(ii) facilitate safe operation, inspection, testing, and maintenance.

314–2 Separate circuits shall be provided for parts of the installation which need to be separately controlled, in such a way that these circuits are not affected by failure of other circuits.

314–3 The number of final circuits required, and the number of points supplied by any final circuit, shall be such as to comply with the requirements of Chapter 43 for overcurrent protection, Chapter 46 for isolation and switching, and Chapter 52 as regards current-carrying capacities of conductors. Standard arrangements of final circuits are described in Appendix 5 but other arrangements are not precluded where these are specified by a suitably qualified electrical engineer.

314–4 Where an installation comprises more than one final circuit, each final circuit shall be connected to a separate way in a distribution board. The wiring of each final circuit shall be electrically separate from that of every other final circuit, so as to prevent indirect energisation of a final circuit intended to be isolated.

CHAPTER 32

EXTERNAL INFLUENCES

NOTE – No Chapter 32 appears in this edition of these Regulations and the chapter number is reserved for future use.

The term 'external influences' signifies influences from sources external to the installation which may affect the design and safe operation of the installation. These external influences fall into three main categories:

A – Environmental conditions,

B – Type of utilisation of premises,

C – Type of building construction.

IEC Publication 364 – 'Electrical installations of buildings', the general arrangement of which has been adopted for this edition of these Regulations, includes as Chapter 32 a classification of external influences, which is reproduced for information as Appendix 6 to these Regulations. This classification will ultimately enable particular types of installation and location to be identified in terms of the significant external influences which distinguish them, and the application of the general requirements will then be qualified in relation to the classes of external influence.

At the time of issue of this edition of these Regulations, the work on requirements for application of the classification of external influences for IEC Publication 364 is insufficiently advanced for adoption as a basis for national regulations. At a later stage, the IEC classification corresponding to Appendix 6 will be considered for adoption as Chapter 32 of these Regulations, together with the proposed IEC requirements for its application.

Some conditions of external influence included in the IEC 364 classification (for example, AC – Altitude) may not be relevant to installations in the United Kingdom. Consideration will be given to excluding these when the classification is adopted for these Regulations.

CHAPTER 33

COMPATIBILITY

331–1 An assessment shall be made of any characteristics of equipment likely to have harmful effects upon other electrical equipment or other services, or likely to impair the supply. These characteristics include, for example:

(i) transient overvoltages,

(ii) rapidly fluctuating loads,

(iii) starting currents,

(iv) harmonic currents (such as with fluorescent lighting loads and thyristor drives),

(v) mutual inductance,

(vi) d.c. feedback,

(vii) high frequency oscillations,

(viii) earth leakage currents,

(ix) any need for additional connections to Earth (e.g. for equipment needing a connection with Earth independent of the main means of earthing of the installation, for the avoidance of interference with its operation).

NOTE – For an external source of energy it is essential that the supply undertaking be consulted regarding any equipment of the installation having a characteristic likely to have a significant influence on the supply, e.g. having heavy starting currents.

CHAPTER 34

MAINTAINABILITY

341–1 An assessment shall be made of the frequency and quality of maintenance that the installation can reasonably be expected to receive during its intended life. This assessment shall, wherever practicable, include consultation with the person or body who will be responsible for the operation and maintenance of the installation. Having regard to the frequency and quality of maintenance expected, the requirements of Parts 4 to 6 of these Regulations shall be applied so that–

(i) any periodic inspection, testing, maintenance and repairs likely to be necessary during the intended life can be readily and safely carried out, and

(ii) the protective measures for safety remain effective during the intended life, and

(iii) the reliability of equipment is appropriate to the intended life.

PART 4

Protection for safety

CONTENTS

400–1 General

CHAPTER 41 — PROTECTION AGAINST ELECTRIC SHOCK

410 Introduction

411 Protection against both direct and indirect contact

412 Protection against direct contact

413 Protection against indirect contact

CHAPTER 42 — PROTECTION AGAINST THERMAL EFFECTS

421 General

422 Protection against fire

423 Protection against burns

CHAPTER 43 — PROTECTION AGAINST OVERCURRENT

431 General

432 Nature of protective devices

433 Protection against overload current

434 Protection against short circuit current

435 Co-ordination of overload and short circuit protection

436 Limitation of overcurrent by characteristics of supply

CHAPTER 44 — *(Reserved for future use)*

CHAPTER 45 — *(Reserved for future use)*

CHAPTER 46 — ISOLATION AND SWITCHING

460 General

461 Isolation

462 Switching off for mechanical maintenance

463 Emergency switching

CHAPTER 47 — APPLICATION OF PROTECTIVE MEASURES FOR SAFETY

470 General

471 Protection against electric shock

472 *(Reserved for future use)*

473 Protection against overcurrent

474 *(Reserved for future use)*

475 *(Reserved for future use)*

476 Isolation and switching

CHAPTER 41

PROTECTION AGAINST ELECTRIC SHOCK

CONTENTS

PART 4

Protection for safety

400–1 General

Chapters 41 to 46 specify requirements for protection of persons, livestock and property and Chapter 47 deals with the application and co-ordination of those requirements.

NOTES: 1 – Protective measures may be applicable to an entire installation, a part, or an item of equipment.

2 – The order in which the protective measures are specified does not imply any relative importance.

CHAPTER 41

PROTECTION AGAINST ELECTRIC SHOCK

410 INTRODUCTION

410–1 Protection against electric shock shall be provided by the application, in accordance with Section 471, of –

an appropriate measure specified in Section 411, for protection against both direct contact (electric shock in normal service) and indirect contact (electric shock in case of a fault), or

a combination of appropriate measures specified in Section 412 for protection against direct contact and Section 413 for protection against indirect contact.

411 PROTECTION AGAINST BOTH DIRECT AND INDIRECT CONTACT

General

411–1 One of the following basic protective measures for protection both against direct contact and indirect contact shall be used, as specified in Section 471:

(i) Protection by safety extra-low voltage (Regulations 411–2 to 411–10)

(ii) Protection by functional extra-low voltage (Regulations 411–11 to 411–15)

(iii) Protection by limitation of discharge of energy (Regulation 411–16)

Protection by safety extra-low voltage

411–2 Protection against electric shock is provided when all the following requirements are fulfilled:

(i) The nominal voltage of the circuit concerned does not exceed extra-low voltage.

(ii) The supply is from one of the safety sources listed in Regulation 411–3.

(iii) The conditions of Regulations 411–4 to 411–10 are fulfilled.

NOTE – Lower voltage limits may be required for certain conditions of external influences (see Section 471).

Safety sources

411–3 The safety source shall be one of the following:

(i) A Class II safety isolating transformer complying with BS 3535, the secondary winding being isolated from Earth.

(ii) A source of current providing a degree of safety equivalent to that of the safety isolating transformer specified in (i) above (e.g. a motor generator with windings providing equivalent isolation).

(iii) An electrochemical source (e.g. a battery) or another source independent of a higher voltage circuit (e.g. a diesel-driven generator).

(iv) Electronic devices where measures have been taken so that even in the case of an internal fault the voltage at the outgoing terminals cannot exceed extra-low voltage.

Arrangement of circuits

411–4 Live parts of safety extra-low voltage circuits shall not be connected to Earth or to live parts or protective conductors forming part of other circuits.

411–5 Exposed conductive parts of safety extra-low voltage circuits shall not intentionally be connected to any of the following:

(i) Earth.

(ii) Protective conductors or exposed conductive parts of another system.

(iii) Extraneous conductive parts, except that where electrical equipment is inherently required to be connected to extraneous conductive parts it shall be verified that those parts cannot attain a voltage exceeding the limit of the safety extra-low voltage circuit. If the exposed conductive parts of safety extra-low voltage circuits are liable to come into contact fortuitously with exposed conductive parts of other circuits, the protection no longer depends solely on the measure for protection by safety extra-low voltage and shall be in accordance with the requirements for the measures applicable to the latter exposed conductive parts.

NOTE – In a consumer's installation required to comply with the Electricity (Factories Act) Special Regulations, 1908 and 1944, the current-using equipment being supplied is required to be of Class II construction.

411–6 Live parts of safety extra-low voltage equipment other than cables shall be electrically separate from those of higher voltage circuits. The electrical separation between live parts of safety extra-low voltage circuits and higher voltage circuits shall be not less than that between the input and output windings of safety isolating transformers.

411–7 Safety extra-low voltage circuit conductors shall preferably be physically separated from those of any other circuit. Where this requirement is impracticable, one of the following arrangements is required:

(i) Safety extra-low voltage circuit conductors shall be insulated in accordance with the requirements of these Regulations for the highest voltage present.

(ii) Safety extra-low voltage circuit cables shall be non-metallic sheathed cables complying with Chapter 52.

(iii) Conductors of circuits at different voltages shall be separated from those at safety extra-low voltage by an earthed metallic screen or an earthed metallic sheath.

(iv) Circuits at different voltages may be contained in a multicore cable or other grouping of conductors but the conductors of safety extra-low voltage circuits shall be insulated, individually or collectively, for the highest voltage present.

NOTE – In arrangements (ii) and (iii) basic insulation of any conductor need be sufficient only for the voltage of the circuit of which it is a part.

411–8 Plugs and socket outlets of safety extra-low voltage circuits shall comply with all of the following requirements:

(i) The plugs shall not be capable of entering socket outlets of other voltage systems in use in the same premises.

(ii) The socket outlets shall exclude plugs of other voltage systems in use in the same premises.

(iii) The socket outlets shall not have a protective conductor connection.

NOTE – In a consumer's installation required to comply with the Electricity (Factories Act) Special Regulations 1908 and 1944, socket outlets having a protective conductor connection are normally necessary unless the equipment being supplied is of Class II construction.

411–9 Mobile safety sources shall be selected or erected in accordance with Regulations 413–18 to 413–26.

411–10 If the nominal voltage exceeds 25V a.c., r.m.s., 50 Hz, or 60V ripple-free d.c., protection against direct contact shall be provided by one or more of the following:

(i) Barriers or enclosures affording at least the degree of protection IP 2X (See BS 5490).

(ii) Insulation capable of withstanding a test voltage of 500V d.c. for one minute.

If the nominal voltage does not exceed 25V a.c., r.m.s., 50 Hz, or 60V ripple-free d.c., protection against direct contact is not required by these Regulations except where specified in Section 471.

Functional extra-low voltage systems

411–11 If for functional reasons extra-low voltage is used but not all the requirements of Regulations 411–2 to 411–10 regarding safety extra-low voltage are fulfilled, the appropriate measures described in Regulations 411–12 to 411–15 shall be taken in order to ensure protection against electric shock. Systems employing these measures are termed 'functional extra-low voltage systems'.

NOTE – Such conditions may, for example, be encountered in extra-low voltage circuits when one point of the extra-low voltage circuit is connected to Earth or if the circuit contains components (such as transformers, relays, remote-control switches, contactors) insufficiently insulated with respect to circuits at higher voltages.

411–12 If the extra-low voltage system complies with the requirements of Section 411 for safety extra-low voltage except that live or exposed conductive parts are connected to Earth or to the protective conductors of other systems (Regulations 411–4 and 411–5) protection against direct contact shall be provided by one or more of the following:

(i) Enclosures giving protection at least equivalent to IP 2X (see BS 5490).
(ii) Insulation capable of resisting a test voltage of 500V r.m.s. for one minute.

Such a system is considered to afford protection against indirect contact.

This requirement does not exclude the installation or the use without supplementary protection of equipment conforming to the relevant British Standard, e.g. BS 415, providing an equivalent degree of safety.

411–13 If the extra-low voltage system does not generally comply with the requirements of Section 411 for safety extra-low voltage, protection against direct contact shall be provided by one or more of the following:

(i) Barriers or enclosures according to Regulations 412–3 to 412–6.

(ii) Insulation corresponding to the minimum test voltage required for the primary circuit.

In addition, protection against indirect contact shall be provided in accordance with Regulation 411–14.

The extra-low voltage circuit may be used to supply factory-built equipment whose insulation does not comply with the minimum test voltage required for the primary circuit provided that the accessible insulation of that equipment is reinforced during erection to withstand a test voltage of 1500V r.m.s. for one minute.

411–14 If the primary circuit of the functional extra-low voltage source is protected by automatic disconnection, exposed conductive parts of equipment in the functional extra-low voltage circuit shall be connected to the protective conductor of the primary circuit.

NOTE – Regulation 411–14 does not exclude the possibility of connecting a conductor of the functional extra-low voltage circuit to the protective conductor of the primary circuit.

If the primary circuit of the functional extra-low voltage source is protected by electrical separation (Regulations 413–35 to 413–39), the exposed conductive parts of equipment in the functional extra-low voltage circuit shall be connected to the non-earthed protective conductor of the primary circuit.

NOTE – This latter requirement does not contravene Regulation 413–37, the combination of the electrically separated circuit and the extra-low voltage circuit being regarded as one electrically separated circuit.

411–15 The socket outlets of functional extra-low voltage systems shall not admit plugs intended for use with other systems.

Protection by limitation of discharge of energy

411–16 For equipment complying with the appropriate British Standard, protection against electric shock is afforded when the equipment incorporates means of limiting the current which can pass through the body of a person or livestock to a value lower than the shock current. Circuits relying on this protective measure shall be separated from other circuits in a manner similar to that specified in Regulations 411–6 and 411–7 for safety extra-low voltage circuits.

412 PROTECTION AGAINST DIRECT CONTACT

General

412–1 One or more of the following basic protective measures for protection against direct contact shall be used, as specified in Section 471:

(i) Protection by insulation of live parts (Regulation 412–2)

(ii) Protection by barriers or enclosures (Regulations 412–3 to 412–6)

(iii) Protection by obstacles (Regulations 412–7 and 412–8)

(iv) Protection by placing out of reach (Regulations 412–9 to 412–13)

Protection by insulation of live parts

412–2 Live parts shall be completely covered with insulation which can only be removed by destruction and which is capable of durably withstanding the mechanical, electrical, thermal and chemical stresses to which it may be subjected in service.

NOTES: 1 – Where insulation is applied during the erection of the installation, the quality of the insulation should be confirmed by tests similar to those which ensure the quality of insulation of similar factory-built equipment.

2 – Paints, varnishes, lacquers and similar products without additional insulation do not generally provide adequate insulation for protection against direct contact.

Protection by barriers or enclosures

412–3 Live parts shall be inside enclosures or behind barriers providing at least the degree of protection IP 2X (see BS 5490) except that, where an opening larger than that admitted for IP 2X is necessary to allow the replacement of parts or to avoid interference with the proper functioning of electrical equipment both the following requirements apply:

(i) Suitable precautions shall be taken to prevent persons or livestock from unintentionally touching live parts.

(ii) It shall be established as far as practicable, that persons will be aware that live parts can be touched through the opening and should not be touched.

NOTE – In consumer's installations required to comply with the Electricity (Factories Act) Special Regulations, 1908 and 1944, more stringent requirements may be applicable.

412–4 Horizontal top surfaces of barriers or enclosures which are readily accessible shall provide a degree of protection of at least IP 4X (see BS 5490).

412–5 Barriers and enclosures shall be firmly secured in place and have sufficient stability and durability to maintain the required degrees of protection and appropriate separation from live parts in the known conditions of normal service.

412–6 Where it is necessary to remove barriers or to open enclosures or to remove parts of enclosures, one or more of the following requirements shall be satisfied:

(i) The removal or opening shall be possible only by use of a key or tool.

(ii) The removal or opening shall be possible only after disconnection of the supply to live parts against which the barriers or enclosures afford protection, restoration of the supply being possible only after replacement or reclosure of the barriers or enclosures.

(iii) An intermediate barrier shall be provided to prevent contact with live parts, such barrier affording a degree of protection of at least IP 2X and removable only by the use of a tool.

This regulation does not apply to ceiling roses complying with BS 67.

Protection by obstacles

412–7 Obstacles shall prevent, as appropriate, either of the following:

(i) Unintentional bodily approach to live parts.

(ii) Unintentional contact with live parts when operating equipment live in normal use.

412–8 Obstacles shall be so secured as to prevent unintentional removal but may be removable without using a key or tool.

Protection by placing out of reach

412–9 Bare or p.v.c.-covered overhead lines for distribution between buildings and structures shall be installed in accordance with the Overhead Line Regulations* (see also Regulation 529–1 and Appendix 11).

412–10 Bare live parts other than overhead lines shall not be within arm's reach.

The term 'arm's reach' refers to a zone of accessibility as described by Figure 1.

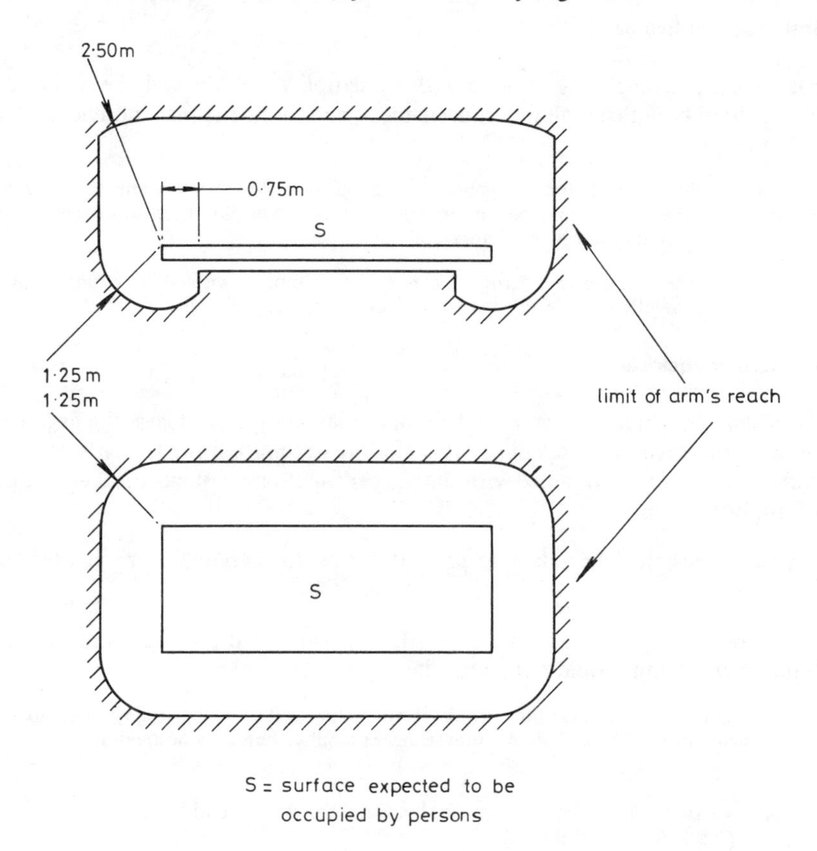

Figure 1.

412–11 Where bare live parts other than overhead lines are out of arm's reach but nevertheless may be accessible, they shall not be within 2.5m of any of the following:

(i) Exposed conductive parts.

(ii) Extraneous conductive parts.

(iii) Bare live parts of other circuits.

412–12 If a normally occupied position is restricted in the horizontal plane by an obstacle (e.g. handrail, mesh screen) affording a degree of protection less than IP 2X, arm's reach shall extend from that obstacle. In the overhead direction, arm's reach is 2.5m from the surface S not taking into account any intermediate obstacle providing a degree of protection less than IP 2X.

NOTE – The values of arm's reach refer to bare hands without any assistance, e.g. from tools or a ladder.

*H.M. Stationery Office

412–13 In places where bulky or long conducting objects are normally handled, the distances required by Regulations 412–10 to 412–12 shall be increased accordingly.

413 PROTECTION AGAINST INDIRECT CONTACT

General

413–1 One or more of the following basic protective measures for protection against indirect contact shall be used, as specified in Section 471:

(i) Earthed equipotential bonding and automatic disconnection of supply (Regulations 413–2 to 413–17)

(ii) Use of Class II equipment or equivalent insulation (Regulations 413–18 to 413–26)

(iii) Non-conducting location (Regulations 413–27 to 413–31)

(iv) Earth-free local equipotential bonding (Regulations 413–32 to 413–34)

(v) Electrical separation (Regulations 413–35 to 413–39)

Protection by earthed equipotential bonding and automatic disconnection of supply

General

413–2 In each installation main equipotential bonding conductors complying with Section 547 shall connect to the main earthing terminal for that installation extraneous conductive parts including the following:

– main water pipes,
– main gas pipes,
– other service pipes and ducting,
– risers of central heating and air conditioning systems,
– exposed metallic parts of the building structure.

NOTES: 1 – This bonding is intended to create a zone in which any voltages between exposed conductive parts and extraneous conductive parts are minimised.

2 – Compliance with Regulation 413–2 will normally satisfy the relevant requirements of the Protective Multiple Earthing Approval, 1974, administered by the Department of Energy.

3 – Bonding to the metalwork of other services may require the permission of the undertakings responsible. Bonding to Post Office earth wires should be avoided unless authorised by the Post Office.

4 – Additional equipotential bonding may be required (see Regulation 413–7 and Section 471).

413–3 The characteristics of the protective devices for automatic disconnection, the earthing arrangements for the installation and the relevant impedances of the circuits concerned shall be co-ordinated so that during an earth fault the voltages between simultaneously accessible exposed and extraneous conductive parts occurring anywhere in the installation shall be of such magnitude and duration as not to cause danger.

NOTES: 1 – For information on types of earthing arrangements, see Appendix 3.

2 – For the conditions of connection of exposed conductive parts, see Regulations 413–8 and 413–9, 413–10 to 413–12 or 413–13 to 413–17 as appropriate to the earthing arrangement concerned.

413–4 Regulation 413–3 is considered to be satisfied if –

(i) for final circuits supplying socket outlets, the earth fault loop impedance at every socket outlet is such that disconnection occurs within 0.4 second. This requirement does not apply to the reduced voltage circuits described in Regulations 471–28 to 471–33. The use of other methods of satisfying Regulation 413–3 is not precluded (see Appendix 7).

(ii) for final circuits supplying only fixed equipment, the earth fault loop impedance at every point of utilisation is such that disconnection occurs within 5 seconds.

413–5 Where protection is afforded by an overcurrent protective device, and the nominal voltage to Earth (U_o) is 240V r.m.s. a.c., the earth fault loop impedance (Z_s), for compliance with Regulation 413–4, shall not exceed the value in Table 41A1 or Table 41A2 as appropriate to the type of circuit and the type and rated current of protective device. For types and rated currents of overcurrent protective device other than those mentioned in the tables, the necessary time/current characteristics shall be obtained from Appendix 8 or from the manufacturer.

NOTE – The tabulated values of impedance in Tables 41A1 and 41A2 are based on the assumption that the earth fault is of negligible impedance.

TABLE 41A1

Maximum earth fault loop impedance (Z_s) for socket outlet circuits

(a) Fuses to BS 88 Part 2

Rating (amperes)	6	10	16	20	25	32	40	50
Z_s (ohms)	8.7	5.3	2.8	1.8	1.5	1.1	0.8	0.6

(b) Fuses to BS 1361

Rating (amperes)	5	15	20	30	45
Z_s (ohms)	11.4	3.4	1.8	1.2	0.6

(c) Fuses to BS 3036

Rating (amperes)	5	15	20	30	45
Z_s (ohms)	9.6	2.7	1.8	1.1	0.6

(d) Fuse to BS 1362

Rating (amperes)	13
Z_s (ohms)	2.5

(e) Type 1 miniature circuit breakers to BS 3871

Rating (amperes)	5	10	15	20	30	50	I_n
Z_s (ohms)	12	6	4	3	2	1.2	$60/I_n$

(f) Type 2 miniature circuit breakers to BS 3871

Rating (amperes)	5	10	15	20	30	50	I_n
Z_s (ohms)	6.8	3.4	2.3	1.7	1.1	0.68	$34/I_n$

(g) Type 3 miniature circuit breakers to BS 3871

Rating (amperes)	5	10	15	20	30	50	I_n
Z_s (ohms)	4.8	2.4	1.6	1.2	0.8	0.48	$24/I_n$

NOTE – When U_o, the nominal voltage to Earth, is other than 240V the tabulated impedance values are to be multiplied by $U_o/240$.

TABLE 41A2

Maximum earth fault loop impedance (Z_s) for circuits supplying fixed equipment

(a) Fuses to BS 88 Part 2

Rating (amperes)	6	10	16	20	25	32	40	50
Z_s (ohms)	13.0	7.7	4.4	3.0	2.4	1.8	1.4	1.1

Rating (amperes)	63	80	100	125	160	200	250
Z_s (ohms)	0.86	0.6	0.45	0.34	0.27	0.19	0.16

Rating (amperes)	315	400	500	630	800
Z_s (ohms)	0.11	0.096	0.065	0.054	0.034

(b) Fuses to BS 1361

Rating (amperes)	5	15	20	30	45	60	80	100
Z_s (ohms)	17	5.3	2.9	2.0	1.0	0.6	0.48	0.28

(c) Fuses to BS 3036

Rating (amperes)	5	15	20	30	45	60	100
Z_s (ohms)	20	5.6	4.0	3.2	1.6	1.2	0.55

(d) Type 1 miniature circuit breakers to BS 3871

Rating (amperes)	5	10	15	20	30	50	I_n
Z_s (ohms)	12	6	4	3	2	1.2	$60/I_n$

(e) Type 2 miniature circuit breakers to BS 3871

Rating (amperes)	5	10	15	20	30	50	I_n
Z_s (ohms)	6.8	3.4	2.3	1.7	1.1	0.68	$34/I_n$

(f) Type 3 miniature circuit breakers to BS 3871

Rating (amperes)	5	10	15	20	30	50	I_n
Z_s (ohms)	4.8	2.4	1.6	1.2	0.8	0.48	$24/I_n$

NOTE – When U_o, the nominal voltage to Earth, is other than 240V the tabulated impedance values are to be multiplied by $U_o/240$.

413–6 Where compliance with the disconnection times of Regulation 413–4 is afforded by a residual current device in an installation which is part of a TN or TT system, the product of the rated residual operating current in amperes and the earth fault loop impedance in ohms shall not exceed 50. If a fault-voltage operated device is used in a TT or IT system the earth fault loop impedance, including the earth electrode resistance, shall not exceed 500 ohms.

NOTE – For TT systems the earth fault loop impedance includes the earth electrode resistance.

413–7 Within the zone formed by the main equipotential bonding, local supplementary bonding connections shall be made to metal parts, to maintain the equipotential zone, where those parts –

(i) are extraneous conductive parts (see definition), and

(ii) are simultaneously accessible with exposed conductive parts or other extraneous conductive parts, and

(iii) are not electrically connected to the main equipotential bonding by permanent and reliable metal-to-metal joints of negligible impedance.

NOTE – Where local equipotential bonding is provided in accordance with Regulation 413–7, metalwork which may be required to be bonded includes baths and exposed metal pipes, sinks, taps, tanks, and radiators and, where practicable, accessible structural metalwork.

Installations which are part of a TN system

413–8 All exposed conductive parts of the installation shall be connected by protective conductors to the main earthing terminal of the installation and that terminal shall be connected to the earthed point of the supply source in accordance with Regulation 542–2, 542–3 or 542–5, as appropriate.

413–9 The protective devices shall be of one or more of the following types:

– overcurrent protective devices,
– residual current devices.

provided that where the neutral and the protective functions are combined in one conductor (PEN conductor) a residual current device shall not be used.

Installations which are part of a TT system

413–10 Where protection is afforded by overcurrent protective devices or residual current devices, exposed conductive parts shall be connected by protective conductors individually, in groups or collectively to an earth electrode or electrodes.

413–11 Where protection is afforded by fault-voltage operated protective devices, all exposed conductive parts and associated extraneous conductive parts protected by any one such protective device shall be connected by protective conductors to an earth electrode via the voltage-sensitive element of that device.

413–12 The protective devices shall be of one or more of the following types:

– residual current devices,
– overcurrent protective devices,
– fault-voltage operated protective devices.

NOTE – Residual current devices are preferred.

Installations which are part of an IT system

413–13 No live conductor of the installation shall be directly connected to Earth.

NOTE – To reduce overvoltage or to damp voltage oscillation, it may be necessary to provide earthing through impedances or artificial neutral points and the characteristics of these should be appropriate to the requirements of the installation.

413–14 Exposed conductive parts shall be earthed either individually, in groups or collectively. Simultaneously accessible exposed conductive parts and associated extraneous conductive parts shall be connected only to an earth electrode common to those parts.

413–15 The protective devices shall be of one or more of the following types:

– residual current devices,
– fault-voltage operated protective devices.

413–16 An insulation monitoring device shall be provided to indicate the occurrence of a first fault from a live part to exposed conductive parts or to Earth. The device shall automatically disconnect the supply, or initiate an audible and/or visual signal.

NOTE – A first fault should be eliminated as quickly as practicable.

413–17 After the occurrence of a first fault, conditions for disconnection of supply, as specified for TN and TT systems, shall apply.

Protection by use of Class II equipment or by equivalent insulation

413–18 Protection shall be provided by one or more of the following:

(i) Electrical equipment of the following types, type tested and marked to the relevant standards:

– electrical equipment having double or reinforced insulation (Class II equipment)
– factory-built assemblies of electrical equipment having total insulation (See BS 5486).

(ii) Supplementary insulation applied to electrical equipment having basic insulation only, as a process in the erection of an electrical installation, providing a degree of safety equivalent to that of electrical equipment according to item (i) above and complying with Regulations 413–20 to 413–26.

(iii) Reinforced insulation applied to uninsulated live parts, as a process in the erection of an electrical installation, providing a degree of safety equivalent to electrical equipment according to item (i) above and complying with Regulations 413–20 to 413–26, such insulation being recognised only where constructional features prevent the application of double insulation.

413–19 The installation of equipment described in item (i) of Regulation 413–18 (for example the fixing and connection of conductors) shall be effected in such a way as not to impair the protection afforded in compliance with the equipment specification.

413–20 The electrical equipment being ready for operation, all conductive parts separated from live parts by basic insulation only shall be contained in an insulating enclosure affording at least the degree of protection IP 2X (see BS 5490).

413–21 The insulating enclosure shall be capable of resisting the mechanical, electrical and thermal stresses to which it is likely to be subjected.

NOTE – Coatings of paint, varnish, and similar products are generally considered not to comply with Regulation 413–21.

413–22 If the insulating enclosure has not previously been tested, a suitable test shall be carried out (see Regulation 613–10).

413–23 The insulating enclosure shall not be pierced by conductive parts, other than circuit conductors, likely to transmit a potential. The insulating enclosure shall not contain any screws of insulating material, the replacement of which by metallic screws could impair the insulation provided by the enclosure.

NOTE – Where the insulating enclosure must be pierced by mechanical joints or connections (e.g. for operating handles of built-in equipment, and for fixing screws) these should be arranged in such a way that protection against indirect contact is not impaired.

413–24 Where lids or doors in the insulating enclosure can be opened without the use of a tool or key, all conductive parts which are accessible if the lid or door is open shall be behind an insulating barrier which prevents persons from coming into contact with those parts; this insulating barrier shall provide a degree of protection of at least IP 2X and be removable only by use of a tool.

413–25 Conductive parts enclosed in the insulating enclosure shall not be connected to a protective conductor. However, provision may be made for connecting protective conductors which necessarily run through the enclosure in order to serve other items of electrical equipment whose supply circuit also runs through the enclosure. Inside that enclosure, any such conductors and their terminals or joints shall be insulated as though they were live parts and their terminals shall be appropriately marked.

413–26 The enclosure provided for this measure shall not adversely affect the operation of the equipment protected.

Protection by non-conducting location

413–27 Exposed conductive parts shall be arranged so that under ordinary circumstances a person will not come into simultaneous contact with –

– two exposed conductive parts, or
– with an exposed conductive part and any extraneous conductive part,

if these parts are liable to be at different potentials through failure of the basic insulation of live parts.

NOTE – Regulation 413–27 is met if the location has an insulating floor and walls, and the distance between the parts mentioned exceeds 2m.

413–28 In a non-conducting location there shall be no protective conductors, and any socket outlets shall not incorporate an earthing contact.

413–29 The resistance of insulating floors and walls at every point of measurement under the conditions specified in Section 613 shall be not less than:

– 50 kΩ where the supply voltage does not exceed 500V, or
– 100 kΩ where the supply voltage exceeds 500V but does not exceed low voltage.

If at any point the resistance is less than the specified value, the floors and walls are extraneous conductive parts for the purposes of protection against shock.

NOTE – Steps may need to be taken so that humidity will not affect the resistance of floors and walls to such an extent that they do not comply with Regulation 413–29.

413–30 The arrangements made shall be permanent. They shall also afford protection where the use of mobile or portable equipment is envisaged.

NOTE – Attention is drawn to the risk that where electrical installations are not under effective supervision, further conductive parts may be introduced at a later date (e.g. mobile or portable Class I equipment or metallic water pipes), which may invalidate compliance with Regulation 413–30.

413–31 Precautions shall be taken so that a potential on extraneous conductive parts in the location cannot be transmitted outside that location.

Protection by earth free local equipotential bonding

413–32 Equipotential bonding conductors shall connect together all simultaneously accessible exposed conductive parts and extraneous conductive parts.

413–33 The local equipotential bonding conductors shall not be in electrical contact with Earth directly, or through exposed conductive parts or through extraneous conductive parts.

NOTE – Where Regulation 413–33 cannot be observed, requirements for protection by automatic disconnection of supply are applicable (see Regulations 413–2 to 413–17).

413–34 Precautions shall be taken so that persons entering the equipotential location cannot be exposed to dangerous potential difference, in particular, where a conductive floor insulated from Earth is connected to the earth-free equipotential bonding conductors.

Protection by electrical separation

413–35 Protection by electrical separation shall be afforded by compliance with Regulations 413–36 and 413–37 and with –

– Regulation 413–38 for a supply to one item of equipment, or
– Regulation 413–39 for a supply to more than one item of equipment.

413–36 The source of supply to the circuit shall comply with the following requirements:

(i) It shall be either:

– a safety isolating transformer complying with BS 3535, the secondary winding being isolated from Earth, or

– a source of current providing a degree of safety equivalent to that of the safety isolating transformer referred to above (e.g., a motor generator with windings providing equivalent isolation).

(ii) Mobile sources of supply fed from a fixed installation shall be selected or installed in accordance with Regulations 413–18 to 413–26.

(iii) Equipment used as a fixed source of supply shall be either:

– selected and installed in accordance with Regulations 413–18 to 413–26, or

– such that the output is separated from the input and from the enclosure by an isolation satisfying the conditions of Regulations 413–18 to 413–26. If such a source supplies several items of equipment, exposed metalwork of that equipment shall not be connected to the metallic enclosure of the source.

(iv) the voltage of the electrically separated circuit shall not exceed 500V.

413–37 The separated circuit shall comply with the following requirements:

(i) Live parts of the separated circuit shall not be connected at any point to another circuit or to Earth and to avoid the risk of a fault to Earth, particular attention shall be given to the insulation of such parts from Earth, especially for flexible cables and cords.

(ii) Flexible cables and cords shall be visible throughout every part of their length liable to mechanical damage.

(iii) A separate wiring system shall preferably be used for the separated circuit. Alternatively, multicore cables without metallic sheath, or insulated conductors in insulating conduit shall be used, their rated voltage being not less than the highest voltage likely to occur, and each circuit shall be protected against overcurrent.

413–38 For a circuit supplying a single item of equipment, no exposed metalwork of the separated circuit shall be connected intentionally either to the protective conductor or to exposed conductive parts of other circuits.

413–39 If precautions are taken to protect the separated circuit from damage and insulation failure, a source of supply complying with Regulation 413–36 items (i) to (iii) may supply more than one item of equipment provided that all the following requirements are fulfilled:

(i) The exposed metalwork of the separated circuit shall be connected together by non-earthed equipotential bonding conductors. Such conductors shall not be connected to the protective conductors or exposed conductive parts of other circuits or to any extraneous conductive parts.

(ii) All socket outlets shall be provided with protective contacts which shall be connected to the equipotential bonding conductors provided in accordance with item (i).

(iii) All flexible cables shall embody a protective conductor for use as an equipotential bonding conductor.

(iv) It shall be verified that, if two faults to exposed metalwork occur and these are fed by conductors of different polarity, an associated protective device will meet the requirements of Regulation 413–3.

CHAPTER 42

PROTECTION AGAINST THERMAL EFFECTS

CONTENTS

CHAPTER 42

PROTECTION AGAINST THERMAL EFFECTS

NOTE – This chapter does not include requirements for protection against overcurrent, for which see Chapter 43.

421 GENERAL

421–1 Protection against thermal effects caused by fixed electrical equipment shall be provided by the appropriate measures specified in this chapter.

421–2 All switchgear shall be selected and erected in accordance with the requirements of Chapter 53 so as to prevent danger from overheating, arcing, or the scattering of hot particles during operation.

422 PROTECTION AGAINST FIRE

422–1 Fixed equipment shall be selected, located and erected so that its intended heat dissipation is not inhibited and it does not present a fire hazard to adjacent building materials.

422–2 Fixed equipment which in normal operation has a surface temperature exceeding 90°C shall be adequately ventilated and be mounted so that no material constituting a fire hazard is within 300mm above or

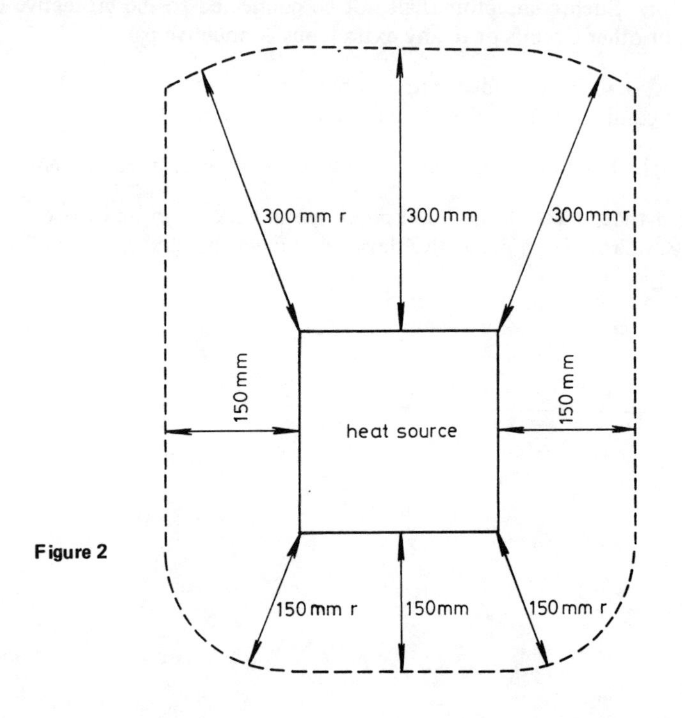

Figure 2

150mm laterally from or below the equipment (see Figure 2), except that these distances may be reduced if a suitable fire-resistant shield or enclosure is installed between the equipment and any such material.

This regulation does not apply to lamps; luminaires complying with BS 4533 and appliances complying with BS 3456 and installed in accordance with the manufacturer's instructions are considered to comply with this regulation.

422–3 Where a distribution board is constructed without a back or without one or more of the other enclosing surfaces it shall be fitted only to surfaces of materials complying with the requirements of BS 5486, Part 13 or to other equipment in such a way as to complete the enclosure.

422–4 All fixed luminaires and lamps shall be placed or guarded so as to prevent ignition of any materials which in the conditions of use foreseen are likely to be placed in proximity to the luminaires or lamps. Any shade or guard used for this purpose shall be suitable to withstand the heat from the luminaire or lamp.

Fixed luminaires mounted in accordance with the manufacturer's installation instructions comply with this regulation.

422–5 Wherever electrical equipment in a single location contains flammable dielectric liquid in excess of 25 litres, precautions shall be taken to prevent burning liquid and the products of combustion of the liquid (flame, smoke, toxic gases) spreading to other parts of the building.

NOTE – Examples of the precautions to be taken are

- a drainage pit to collect leakages of liquid and ensure their extinction in the event of fire, or
- installation of the equipment in a chamber of adequate fire resistance and the provision of sills or other means of preventing burning liquid spreading to other parts of the building, such chamber being ventilated solely to the external atmosphere.

423 PROTECTION AGAINST BURNS

423–1 Where the temperature of any part of the enclosure of fixed equipment is liable to exceed 80°C, the enclosure shall be so located or guarded as to prevent accidental contact therewith. This regulation does not apply to equipment which is the subject of and complies with a British Standard which admits a temperature exceeding 80°C for an unguarded enclosure.

CHAPTER 43

PROTECTION AGAINST OVERCURRENT

CONTENTS

CHAPTER 43

PROTECTION AGAINST OVERCURRENT

431 GENERAL

431–1 Live conductors shall be protected by one or more devices for automatic interruption of the supply in the event of overload (see Section 433) and short circuits (Section 434) except in cases where the overcurrent is limited in accordance with Section 436.

NOTES: 1 – Live conductors protected against overload in accordance with Section 433 are considered to be protected also against faults likely to cause overcurrents of a magnitude similar to overload currents.

2 – Protection of conductors in accordance with this chapter does not necessarily protect the equipment connected to the conductors or flexible cables connecting such equipment to fixed installations through plugs and socket outlets.

431–2 The protection against overload and the protection against short circuits shall be co-ordinated in accordance with Section 435.

432 NATURE OF PROTECTIVE DEVICES

General

432–1 The protective devices shall be of the appropriate types indicated in Regulations 432–2 to 432–4.

Protection against both overload and short circuit currents

432–2 Devices providing protection against both overload current and short circuit current shall be capable of making and breaking any overcurrent up to and including the prospective short circuit current at the point where the device is installed. They shall satisfy the requirements of Section 433 and Regulations 434–4 and 434–5. Such protective devices may be circuit breakers incorporating overload release, or fuses, or circuit breakers in conjunction with fuses.

NOTE – The use of a protective device having a rated breaking capacity below the value of the prospective short circuit current at its place of installation is subject to the requirements of Regulation 434–4.

Protection against overload current only

432–3 Devices providing protection against overload current shall satisfy the requirements of Section 433. Such devices may have a breaking capacity below the value of the prospective short circuit current at the point where the device is installed.

Protection against short circuit current only

432–4 Devices providing protection against short circuit currents shall satisfy the requirements of Section 434. Such devices shall be capable of making and breaking short circuit currents up to and including the prospective short circuit current. Such devices may be circuit breakers with short circuit release, or fuses.

433 PROTECTION AGAINST OVERLOAD CURRENT

General

433–1 Protective devices shall be provided to break any overload current flowing in the circuit conductors before such a current could cause a temperature rise detrimental to insulation, joints, terminations, or surroundings of the conductors.

Co-ordination between conductors and protective devices

433–2 The characteristics of a device protecting a circuit against overload shall satisfy the following conditions:

(i) its nominal current or current setting (I_n) is not less than the design current (I_B) of the circuit, and

(ii) its nominal current or current setting (I_n) does not exceed the lowest of the current-carrying capacities (I_z) of any of the conductors of the circuit, and

(iii) the current causing effective operation of the protective device (I_2) does not exceed 1.45 times the lowest of the current-carrying capacities (I_z) of any of the conductors of the circuit.

Where the device is a fuse to BS 88 or BS 1361 or a circuit breaker to BS 3871 Part 1 or BS 4752 Part 1 compliance with condition (ii) also results in compliance with condition (iii).

Where the device is a semi-enclosed fuse to BS 3036 compliance with condition (iii) is afforded if its nominal current (I_n) does not exceed 0.725 times the current-carrying capacity of the lowest rated conductor in the circuit protected.

NOTES: 1 – The conditions of Regulation 433–2 may be stated as formulae as follows:

$$I_B \leqslant I_n \leqslant I_z$$

$$I_2 \leqslant 1.45 . I_z$$

2 – It is expected that the circuit is so designed that small overloads of long duration will not frequently occur (see Section 311).

3 – For current-carrying capacities of cables, see Appendix 9.

Protection of conductors in parallel

433–3 When the same protective device protects conductors in parallel, the value of I_z is the sum of the current-carrying capacities of those conductors. This provision is applicable only if those conductors are of the same type, cross-sectional area, length and disposition, have no branch circuits throughout their length and are arranged so as to carry substantially equal currents. This regulation does not apply to ring circuits.

434 PROTECTION AGAINST SHORT CIRCUIT CURRENT

NOTE – Section 434 deals only with the case of short circuits anticipated between conductors belonging to the same circuit.

General

434–1 Protective devices shall be provided to break any short circuit current in the conductors of each circuit before such current could cause danger due to thermal and mechanical effects produced in conductors and connections. The nominal current of such a protective device may be greater than the current-carrying capacity of the conductor being protected.

Determination of prospective short circuit currents

434–2 The prospective short circuit current at every relevant point of the complete installation shall be determined. This may be done either by calculation or by measurement of the relevant impedances.

NOTE – If the prospective short circuit current at the origin of an installation is less than the breaking capacity rating of the smallest rated protective device to be used in the installation, no further assessment of the prospective short circuit current is necessary.

Characteristics of short circuit protective devices

434–3 Each short circuit protective device shall meet the requirements of Regulation 434–4 and Regulation 434–5.

434–4 The breaking capacity rating shall be not less than the prospective short circuit current at the point at which the device is installed, except where the following paragraph applies.

A lower breaking capacity is permitted if another protective device having the necessary breaking capacity is installed on the supply side. In that case the characteristics of the devices shall be co-ordinated so that the energy let-through of these two devices does not exceed that which can be withstood without damage by the device on the load side and the conductors protected by these devices.

Other characteristics may need to be taken into account, such as dynamic stresses and arcing energy, for the device on the load side.

NOTE – Details of the characteristics needing co-ordination should be obtained from the manufacturers of the devices concerned. Co-ordination with regard to their operating times is of importance.

434–5 Where an overload protective device complying with Section 433 is to be used also for short circuit protection, and has a rated breaking capacity not less than the value of the prospective short circuit current at its point of installation, it may generally be assumed that the requirements of these Regulations are satisfied as regards short circuit protection of the conductor on the load side of that point.

NOTE – For certain types of circuit breaker, especially non-current-limiting types, this assumption may not be valid for the whole range of short circuit currents; in case of doubt, its validity should be checked in accordance with the requirements of Regulation 434–6.

434–6 Where Regulation 434–5 does not apply, it shall be verified as follows that all currents caused by a short circuit occurring at any point of the circuit shall be interrupted in a time not exceeding that which brings the cable conductors to the admissible limiting temperature. In addition it shall be verified for larger installations that the cable is unlikely to be damaged mechanically.

For short circuits of duration up to 5s, and for cable conductors of cross-sectional area of 10mm^2 or greater, the time t in which a given short-circuit current will raise the conductors from the highest permissible temperature in normal duty to the limit temperature, can as an approximation be calculated from the formula:

$$t = \frac{k^2 S^2}{I^2}$$

where

t = duration in s,
S = cross-sectional area in mm^2
I = effective short-circuit current in A expressed, for a.c., as the r.m.s. value,
k = 115 for copper conductors insulated with p.v.c.,
134 for copper conductors insulated with 60°C rubber, 85°C rubber,
143 for copper conductors with 90°C thermosetting insulation,
108 for copper conductors insulated with impregnated paper,
135 for mineral-insulated cables with copper conductors,
76 for aluminium conductors insulated with p.v.c.,
89 for aluminium conductors insulated with 60°C rubber, 85°C rubber,
94 for aluminium conductors with 90°C thermosetting insulation,
71 for aluminium conductors insulated with impregnated paper,
87 for mineral-insulated cables with aluminium conductors,
100 for tin-soldered joints in copper conductors, corresponding to a temperature of 160°C.

NOTE – For very short durations (less than 0.1s) where asymmetry of the current is of importance and for current limiting devices, the value of k^2S^2 for the cable should be greater than the value of let-through energy (I^2t) of the device as quoted by the manufacturer.

Protection of conductors in parallel

434–7 A single device may protect several conductors in parallel against short circuit provided that the operating characteristics of the device and the method of installation of the parallel conductors are suitably co-ordinated; for the selection of the protective device, see Chapter 53.

NOTE – Account should be taken of the conditions that would occur in the event of a short circuit which does not affect all of the conductors.

435 CO-ORDINATION OF OVERLOAD AND SHORT CIRCUIT PROTECTION

435–1 The characteristics of devices for overload protection and those for short circuit protection shall be co-ordinated so that the energy let-through by the short circuit protective device does not exceed that which can be withstood without damage by the overload protective device.

NOTE – For circuits incorporating motor starters, Regulation 435–1 does not preclude the type of co-ordination described in BS 4941.

436 LIMITATION OF OVERCURRENT BY CHARACTERISTICS OF SUPPLY

436–1 Conductors are considered to be protected against overload and short circuit currents where they are supplied from a source incapable of supplying a current exceeding the current-carrying capacity of the conductors.

CHAPTER 44

Reserved for future use

CHAPTER 45

Reserved for future use

CHAPTER 46

ISOLATION AND SWITCHING

CONTENTS

CHAPTER 46

ISOLATION AND SWITCHING

460 GENERAL

460–1 Means shall be provided for non-automatic isolation and switching to prevent or remove hazards associated with the electrical installation or electrically powered equipment and machines. These means shall comply with the appropriate requirements of this chapter and of Sections 476 and 537.

460–2 In TN-C systems the PEN conductor shall not incorporate a means of isolation or switching. In TN-S systems the protective conductor shall not incorporate a means of isolation or switching and provision need not be made for isolation of the neutral conductor except where this is specifically required by Section 554.

461 ISOLATION

461–1 Every circuit shall be provided with means of isolation from each of the live supply conductors, except as provided in Regulation 460–2. It is permissible to isolate a group of circuits by a common means, due consideration being given to service conditions.

461–2 Adequate provision shall be made so that precautions can be taken to prevent any equipment from being unintentionally energised.

461–3 Where an item of equipment or enclosure contains live parts that are not capable of being isolated by a single device, a warning notice shall be placed in such a position that any person gaining access to live parts will be warned of the need to use the appropriate isolating devices, unless an interlocking arrangement is provided so that all the circuits concerned are isolated.

461–4 Where necessary to prevent danger, adequate means shall be provided for the discharge of capacitive electrical energy.

461–5 All devices used for isolation shall be clearly identifiable, e.g. by marking, to indicate the circuit which they isolate.

462 SWITCHING OFF FOR MECHANICAL MAINTENANCE

462–1 Means of switching off for mechanical maintenance shall be provided where mechanical maintenance may involve a risk of physical injury.

462–2 Devices for switching off for mechanical maintenance shall be suitably placed, readily identifiable (e.g. by marking if necessary) and convenient for their intended use.

462–3 Suitable means shall be provided so that precautions can be taken to prevent any equipment from being unintentionally or inadvertently reactivated.

463 EMERGENCY SWITCHING

463–1 For every part of an installation which it may be necessary to disconnect rapidly from the supply in order to prevent or remove a hazard, a means of emergency switching shall be provided.

463–2 Means for emergency switching shall act as directly as possible on the appropriate supply conductors, and shall be such that a single initiative action will cut off the appropriate supply.

463–3 The arrangement of emergency switching shall be such that its operation does not introduce a further hazard or interfere with the complete operation necessary to remove the hazard.

463–4 Devices for emergency switching shall be readily accessible and suitably marked.

463–5 Means of emergency stopping shall be provided where movements caused by electrical means may give rise to danger.

CHAPTER 47

APPLICATION OF PROTECTIVE MEASURES FOR SAFETY

CONTENTS

470	GENERAL
471	PROTECTION AGAINST ELECTRIC SHOCK
471–1	General
471–2 to **471–5**	**Protection against both direct and indirect contact**
471–2 and **471–3**	Safety extra-low voltage
471–4	Functional extra-low voltage
471–5	Limitation of discharge of energy
471–6 to **471–10**	**Protection against direct contact**
471–6	Insulation of live parts
471–7 and **471–8**	Barriers or enclosures
471–9	Obstacles
471–10	Placing out of reach
471–11 to **471–21**	**Protection against indirect contact**
471–11 to **471–15**	Automatic disconnection of supply
471–16 to **471–18**	Class II equipment or equivalent insulation
471–19	Non-conducting location
471–20	Earth-free local equipotential bonding
471–21	Electrical separation
471–22 to **471–26**	**Special provisions and exemptions**
471–27 to **471–33**	**Automatic disconnection and reduced system voltages**
471–34 to **471–46**	**Protective measures for particular locations**
471–34 to **471–39**	Bathrooms and showers
471–40 and **471–41**	Agricultural installations
471–42 to**471–46**	Installations for mobile caravans and related sites
472	*(Reserved for future use)*
473	PROTECTION AGAINST OVERCURRENT
473–1 to **473–4**	**Protection against overload**
473–1 and **473–2**	Position of devices for overload protection
473–3	Omission of devices for overload protection

473–4	Overload protective devices in IT systems
473–5 to **473–8**	**Protection against short circuit**
473–5 to **473–7**	Position of devices for short circuit protection
473–8	Omission of devices for short circuit protection
473–9 to **473–14**	**Protection according to the nature of circuits and distribution system**
473–9	Phase conductors
473–10 to **473–12**	Neutral conductor, TN or TT systems
473–13 and **473–14**	Neutral conductor, IT systems
474	*(Reserved for future use)*
475	*(Reserved for future use)*
476	ISOLATION AND SWITCHING
476–1	General
476–2 to **476–6**	Isolation
476–7 and **476–8**	Switching off for mechanical maintenance
476–9 to **476–13**	Emergency switching
476–14 to **476–20**	Other requirements for switching

CHAPTER 47

APPLICATION OF PROTECTIVE MEASURES FOR SAFETY

470 GENERAL

470–1 Protective measures shall be applied in every installation or part of an installation, or to equipment, as required by –

– Section 471 – Protection against electric shock,
– Section 473 – Protection against overcurrent,
– Section 476 – Isolation and switching.

470–2 Precautions shall be taken so that no mutual detrimental influence can occur between various protective measures in the same installation or part of an installation.

471 PROTECTION AGAINST ELECTRIC SHOCK

General

471–1 The application of the various protective measures described in Chapter 41 (relevant regulation numbers indicated below in brackets) is qualified as follows.

Protection against both direct and indirect contact

Safety extra-low voltage (Regulations 411–2 to 411–10)

471–2 Where the use of safety extra-low voltage (SELV) is relied upon for protection against direct contact, i.e. where live parts are not insulated or provided with barriers and enclosures in accordance with Regulation 411–10, the nominal voltage shall not in any event exceed 25V r.m.s. a.c. or 60V ripple-free d.c. These voltage limits are applicable only to conditions where simultaneously accessible parts may be touched by a person having a body resistance assumed as conventionally normal, and shall be appropriately reduced in conditions where reduced or very low body resistance is to be expected.

471–3 Where SELV is used for protection against indirect contact only, and where the live parts of the SELV circuit are insulated or provided with barriers and enclosures in accordance with Regulation 411–10, the nominal voltage shall not in any event exceed 50V r.m.s. a.c., or 120V ripple-free d.c. These voltage limits are applicable only to conditions where simultaneously accessible parts may be touched by a person having a conventionally normal body resistance, and shall be appropriately reduced in conditions where reduced or very low body resistance is to be expected.

NOTE to 471–2 and 471–3 Conventionally normal body resistance relates to a contact involving one hand and both feet, the skin being dry or moist with perspiration (but not wet). Reduced body resistance may be expected in situations where the hands and/or feet are likely to be wet or where the shock current path may not be through the extremities, and very low body resistance (of the order of one quarter of the conventionally normal body resistance) is to be expected in locations where a person is immersed in water or working in confined conductive locations.

Functional extra-low voltage (Regulations 411–11 to 411–15)

471–4 Where, for functional reasons, extra-low voltage is used but –

(i) one point of the extra-low voltage circuit is required to be earthed, or

(ii) live parts or exposed conductive parts of the extra-low voltage circuit are connected to the protective conductors of other systems (whether those protective conductors are earthed or not), or

(iii) the insulation between the extra-low voltage circuits and other circuits is not equivalent to that provided by a safety source,

the system shall be treated as a functional extra-low voltage system and Regulations 411–11 to 411–15 apply.

Limitation of discharge of energy (Regulation 411–16)

471–5 This measure shall be applied only to individual items of current-using equipment complying with an appropriate British Standard, where the equipment incorporates means of limiting to a safe value the current that can flow from the equipment through the body of a person or livestock. The application of this measure may be extended to a part of an installation derived from such items of equipment, where the British Standard concerned provides specifically for this, e.g. to electric fences supplied from electric fence controllers complying with BS 2632.

Protection against direct contact

Insulation of live parts (Regulation 412–2)

471–6 This measure relates to basic insulation, and is intended to prevent contact with live parts. It is generally applicable for protection against direct contact, in conjunction with a measure for protection against indirect contact.

Barriers or enclosures (Regulations 412–3 to 412–6)

471–7 This measure is intended to prevent or deter any contact with live parts. It is generally applicable for protection against direct contact in conjunction with a measure for protection against indirect contact.

471–8 The exception in Regulation 412–3 allowing for openings larger than IP 2X in barriers or enclosures shall be applied only to items of equipment or accessories complying with a British Standard where compliance with the generality of Regulation 412–3 is impracticable by reason of the function of those items, e.g. to lampholders complying with BS 5042 or other appropriate Standard. Wherever that exception is used, the opening shall be as small as is consistent with the requirements for proper functioning and for replacement of parts.

Obstacles (Regulations 412–7 and 412–8)

471–9 This measure is intended to prevent unintentional contact with live parts, but not intentional contact by deliberate circumvention of the obstacles. It shall be used only for protection against direct contact in areas accessible only to skilled persons, or instructed persons under direct supervision.

Placing out of reach (Regulations 412–9 to 412–13)

471–10 This measure is intended only to prevent unintentional contact with live parts and shall be applied only for protection against direct contact. The application of the provisions of Regulations 412–10 to 412–12 shall be limited to locations accessible only to skilled persons, or instructed persons under direct supervision.

Protection against indirect contact

Automatic disconnection of supply (Regulations 413–2 to 413–17)

471–11 This measure is generally applicable, and is intended to prevent the occurrence of voltages of such magnitude and duration between simultaneously accessible conductive parts that danger could arise. It includes all methods involving the earthing of exposed conductive parts. The limiting values of earth fault loop impedance specified in Regulation 413–5 shall be applied where the conditions are such that conventionally normal body resistance applies. In conditions where reduced or very low body resistance is to be expected, either the earth fault loop impedance values shall be appropriately reduced or another protective measure shall be used.

NOTE – Conventionally normal body resistance relates to a contact involving one hand and both feet, the skin being dry or moist with perspiration (but not wet). Reduced body resistance may be expected in situations where the hands and/or feet are likely to be wet or where the shock current path may not be through the extremities, and very low body resistance (of the order of one quarter of the conventionally normal body resistance) is to be expected in locations where a person is immersed in water or working in confined conductive locations.

471–12 The limiting values of earth fault loop impedance specified in Regulation 413–5 are applicable only where the exposed conductive parts of the equipment concerned and any extraneous conductive parts are situated within the zone created by the main equipotential bonding (see also Regulation 413–2).

Where a circuit originating in that zone is specifically intended to supply equipment to be used outside the zone, and that equipment may be touched by a person in contact directly with the general mass of Earth, the following requirements apply:

(i) For equipment supplied by means of a socket outlet rated at 32A or less, or by means of a flexible cable or cord having a similar current-carrying capacity, protection shall be afforded by a residual current device having a rated residual operating current not exceeding 30mA.

(ii) For all other equipment, the earth loop impedance shall be such that disconnection occurs within 0.4 second.

471–13 Where the measure is used in a household or similar installation forming part of a TT system, every socket outlet circuit shall be protected by a residual current device having a rated residual operating current not exceeding 30mA.

471–14 Automatic disconnection using residual current devices shall not be applied to circuits incorporating a PEN conductor. The measure is otherwise generally applicable, provided that the device is selected to have a residual operating current ensuring compliance with Regulation 413–3 and Regulation 413–6. The use of such devices is preferred where the value of earth fault loop impedance prevents the use of overcurrent devices to obtain compliance with the disconnection times specified in Regulation 413–4.

NOTE – If a residual current device affording protection against indirect contact has a rated residual current equal to or less than 30mA and an operating time of 40 ms or less at a residual current of 250 mA, it may also be used to reduce the risk associated with direct contact in case of failure of other protective measures. Such a device cannot be used as a sole means of protection against direct contact and does not obviate the need to apply one of the protective measures specified in items (i) to (iv) of Regulation 412–1.

471–15 Automatic disconnection using fault-voltage operated protective devices is recognised for use in TT and IT systems and is suitable where the impedance of the earth fault loop prevents compliance with Regulations 413–4 and 413–5 by the use of overcurrent protective devices.

NOTE – The use of residual current devices is preferred.

Class II equipment or equivalent insulation (Regulations 413–18 to 413–26)

471–16 This measure is intended to prevent the appearance of a dangerous voltage on the exposed metalwork of electrical equipment through a fault in the basic insulation. It is generally applicable to items of equipment, either by the selection of equipment complying with an appropriate British Standard where that standard provides for the use of Class II construction or total insulation, or by the application of suitable supplementary insulation during erection.

471–17 Where a circuit supplies items of Class II equipment, a means of connection to the protective conductor of the circuit shall nevertheless be provided at every point for the supply of current-using equipment likely to be changed by the user, and at terminations for accessories similarly likely to be changed by the user. This requirement need not be observed where Regulation 471–18 applies. Where a Class II equipment has exposed metalwork connected during erection to a protective conductor, that metalwork shall be treated as exposed conductive parts and the equipment shall be treated as Class I equipment.

471–18 Where this measure is to be used as a sole means of protection against indirect contact (i.e., where a whole installation or circuit is intended to consist entirely of Class II equipment or the equivalent), it shall be verified that the installation or circuit concerned will be under effective supervision in normal use so that no change is made that would impair the effectiveness of the Class II or equivalent insulation. The measure shall not therefore be so applied to any circuit which includes socket outlets or where a user may change items of equipment without authorisation. Cables having a non-metallic sheath or a non-metallic enclosure shall not be described as being of Class II construction. However, the use of such cables installed in accordance with Chapter 52 is considered to afford satisfactory protection against direct and indirect contact.

Non-conducting location (Regulations 413–27 to 413–31)

471–19 This measure is intended to prevent simultaneous contact with parts which may be at different potentials through failure of the basic insulation of live parts. It is not recognised in these Regulations for general use, but may be applied in special situations under effective supervision, where specified by a suitably qualified electrical engineer.

Earth free local equipotential bonding (Regulations 413–32 to 413–34)

471–20 This measure is intended to prevent the appearance of a dangerous voltage between simultaneously accessible parts in the event of failure of the basic insulation. It shall be applied only in special situations which are Earth free and under effective supervision and where specified by a suitably qualified electrical engineer.

Electrical separation (Regulations 413–35 to 413–39)

471–21 This measure is intended, in an individual circuit, to prevent shock currents through contact with exposed conductive parts which might be energised by a fault in the basic insulation of that circuit. It may be applied to the supply of any individual item of equipment by means of a transformer complying with BS 3535 the secondary of which is not earthed, or a source affording equivalent safety. Its use to supply several items of equipment from a single separated source is recognised in these Regulations only for special situations under effective supervision, where specified by a suitably qualified electrical engineer.

Special provisions and exemptions

471–22 For areas to which only skilled persons, or instructed persons under direct supervision, have access it is sufficient to provide against unintentional contact with live parts by the use of obstacles in accordance with Regulations 412–7 and 412–8, or by placing of live parts out of reach in accordance with Regulations 412–9 to 412–13, subject also to Regulations 471–23 to 471–25.

471–23 The dimensions of passage-ways and working platforms for open type switchboards and other equipment having exposed live parts shall comply with Regulation 17 of the Electricity (Factories Act) Special Regulations, as appropriate to the nominal voltage of the live parts.

471–24 For areas which are accessible only to skilled persons by the use of a safety ward lock key or tools, the measures of protection against electric shock specified in Chapter 41 and in this section may be totally dispensed with, where this is permitted by the Electricity (Factories Act) Special Regulations.

471–25 Areas reserved for skilled or instructed persons shall be clearly and visibly indicated by suitable warning signs.

471–26 It is permissible to dispense with measures of protection against indirect contact in the following instances:

(i) Overhead line insulator wall brackets and metal parts connected to them if such parts are not situated within arm's reach.

(ii) Steel reinforced concrete poles in which the steel reinforcement is not accessible.

(iii) Exposed conductive parts which, owing to their reduced dimensions or their disposition cannot be gripped or cannot be contacted by a major surface of the human body, provided that connection of these parts to a protective conductor cannot readily be made or cannot be reliably maintained.

NOTE – This item applies to small isolated metal parts such as bolts, rivets, nameplates and cable clips. For the purposes of Item (iii) a major surface of the human body is considered to be 50mm x 50mm.

(iv) Fixing screws for non-metallic accessories provided that there is no appreciable risk of the screws coming into contact with live parts.

(v) Short lengths of metal conduit for mechanical protection of cables having a non-metallic sheath, or other metal enclosures mechanically protecting equipment complying with Regulations 471–16 to 471–18.

Automatic disconnection and reduced system voltages

471–27 Where for functional reasons the use of extra-low voltage is impracticable and there is no requirement for the use of safety extra-low voltage, a reduced low voltage system may be used as specified in Regulations 471–28 to 471–33.

471–28 The nominal voltage of the reduced low voltage circuits shall not exceed 110V r.m.s. a.c. between phases (three phase 65V to earthed neutral, single phase 55V to earthed midpoint).

471–29 The source of supply to reduced low voltage circuits shall be one of the following:

– a double wound isolating transformer complying with BS 3535, or

– a motor generator having windings providing isolation equivalent to that provided by the windings of an isolating transformer, or

– a source independent of other supplies, e.g. a diesel generator.

471–30 The neutral (star) point of the secondary windings of three-phase transformers and generators, or the midpoint of the secondary windings of single-phase transformers and generators, shall be connected to Earth.

471–31 Protection against direct contact shall be provided by insulation in accordance with Regulation 471–6 or by barriers or enclosures in accordance with Regulations 471–7 and 471–8.

471–32 Protection against indirect contact by automatic disconnection shall be provided by means of an overcurrent protective device in each phase conductor or by a residual current device, and all exposed conductive parts of the reduced low voltage system shall be connected to Earth. The earth fault loop impedance at every point of utilisation, including socket outlets, shall be such that the disconnection time does not exceed 5 seconds. Where a residual current device is used, the product of the rated residual operating current in amperes and the earth fault loop impedance in ohms shall not exceed 50.

471–33 Plugs, socket outlets and cable couplers of reduced low voltage systems shall have a protective conductor contact and shall not be interchangeable with plugs, socket outlets and cable couplers for use at other voltages in the same installation.

Protective measures for particular locations

NOTE – In certain locations the susceptibility of persons, and livestock where present, to electric shock may be so high as to necessitate special combinations of protective measures and supplementary precautions as specified in the following Regulations 471–34 to 471–46.

Bathrooms and showers

471–34 In a room containing a fixed bath or shower, there shall be no socket outlets and there shall be no provision for connecting portable equipment. Where shower cubicles are located in rooms other than bathrooms, any socket outlets shall be situated at least 2.5m from the shower cubicle. These requirements do not apply to shaver supply units complying with Regulation 471–37.

471–35 In a room containing a fixed bath or shower, supplementary equipotential bonding shall be provided between simultaneously accessible exposed conductive parts of equipment, between exposed conductive parts and simultaneously accessible extraneous conductive parts, and between simultaneously accessible extraneous conductive parts (see definition).

471–36 For circuits supplying equipment in a room containing a fixed bath or shower, where the equipment is simultaneously accessible with exposed conductive parts of other equipment or with extraneous conductive parts, the characteristics of the protective devices and the earthing arrangements shall be such that in the event of an earth fault, disconnection occurs within 0.4 seconds.

471–37 In a room containing a fixed bath or shower, electric shavers shall be connected only by means of a shaver supply unit complying with BS 3052. The earthing terminal of the shaver supply unit shall be connected to the protective conductor of the final circuit from which the supply is derived.

471–38 In a room containing a fixed bath or shower cubicle, parts of a lampholder within a distance of 2.5m from the bath or shower cubicle shall be constructed of or shrouded in insulating material. Bayonet-type (B 22) lampholders shall be fitted with a protective shield complying with BS 5042. As an alternative, totally enclosed luminaires may be used.

471–39 Every switch or other means of electrical control or adjustment shall be so situated as to be normally inaccessible to a person using a fixed bath or shower. This requirement does not apply to electric shaver supply units installed in accordance with Regulation 471–37 or to insulating cords of cord-operated switches, or to controls incorporated in instantaneous water heaters complying with the relevant requirements of BS 3456: Section 3.9. No stationary appliance having heating elements which can be touched shall be installed within reach of a person using the bath or shower.

Agricultural installations

471–40 In situations accessible to livestock in and around agricultural buildings, electrical equipment shall so far as is practicable be of Class II construction, or constructed of or protected by suitable insulating material.

Where protection against indirect contact is provided by automatic disconnection in such situations, the limiting values of earth fault loop impedance prescribed in Regulation 413–5 are not applicable and shall be reduced as appropriate to the type of livestock whose presence is envisaged.

NOTE – The very low body resistance of horses and cattle, for example, makes them susceptible to electric shock at voltages lower than 25V r.m.s. a.c.

471–41 Where protection by the use of safety extra-low voltage is used in situations accessible to livestock in and around agricultural buildings, the upper limit of nominal voltage specified in Regulation 411–2 does not apply and shall be reduced as appropriate.

Installations for mobile (touring) caravans and related sites

NOTE – Regulations 471–42 to 471–46 relate only to touring caravans having a maximum demand not exceeding 16A and their sites. For other types of caravan, e.g. those that are designed to be transported, the Regulations generally are applicable. Installations in caravans such as mobile workshops may be subject to the Electricity (Factories Act) Special Regulations 1908 and 1944.

471–42 Installations on sites for the supply of mobile (touring) caravans shall allow for the use of protection against indirect contact by automatic disconnection of the supply.

471–43 The earth terminal of every socket outlet supplying a mobile (touring) caravan shall be earthed by one of the following means:

(i) for a site installation forming part of a TN-S system, connection to the main earthing terminal of that installation by means of a protective conductor having a high degree of reliability, e.g. a protective conductor in a suitable underground cable or a duplicated protective conductor in an overhead system, or

(ii) to an independent earth electrode associated with a residual current protective device. This device may protect a single socket outlet or a group of socket outlets as described in Regulation 471–44, or

(iii) for a site installation forming part of a TN-C-S system, connection to the main earthing terminal of that installation, subject to the formal consent of the Department of Energy.

NOTES: 1 – For item (iii) consultation with the supply undertaking will indicate the requirements for approval and generally the installation will require the use of armoured cables, concentric PEN-conductor cables, or other approved arrangement, with the provision of protective conductors throughout the site distribution. An earth electrode may be required on the distributed protective conductor in addition to connection to the PME terminal.

2 – The main earthing terminal of a site installation forming part of a TN-C-S system may be used for the protection of installations of permanent buildings on the site. Formal consent of the Department of Energy is not required in this case.

471–44 Each mobile (touring) caravan shall be supplied from a socket outlet of the site installation complying with Regulation 553–6. Each such socket outlet shall be supplied, either singly or in a group not exceeding six socket outlets, through a residual current device, the device having a residual operating current not exceeding 30mA and complying with Regulation 413–6.

471–45 The internal installation of every mobile (touring) caravan shall allow for the use of protection by automatic disconnection. A protective conductor shall be provided throughout each circuit of the caravan installation, and socket outlets shall incorporate an earthing contact. This regulation does not exclude the use of Class II equipment.

471–46 Extraneous conductive parts of every mobile (touring) caravan shall be bonded to the protective conductor of the caravan installation, if necessary in more than one place if the type of construction does not provide continuity. The nominal cross-sectional area of equipotential bonding conductors for this purpose shall be not less than $4mm^2$. If the caravan is made substantially of insulating material (e.g. glass-reinforced plastics) these requirements do not apply to isolated metal parts which are unlikely to become live in the event of a fault. (See also Chapter 54 for bonding requirements).

472 *(Reserved for future use)*

473 PROTECTION AGAINST OVERCURRENT

Protection against overload

Position of devices for overload protection

473–1 A device for protection against overload shall be placed at the point where a reduction occurs in the value of current-carrying capacity of the conductors of the installation. This requirement does not apply where the arrangements mentioned in Regulation 473–2 are adopted, and no overload protective device need be provided where Regulation 473–3 applies.

NOTE – A reduction in the value of current-carrying capacity may be caused by a change in cross-sectional area, method of installation, type of cable or conductor, or in environmental conditions.

473–2 The device protecting conductors against overload may be placed along the run of those conductors, provided that the part of run between the point where the value of current-carrying capacity is reduced and the position of the protective device has no branch circuits or outlets for the connection of current-using equipment.

Omission of devices for overload protection

473–3 Devices for protection against overload need not be provided for –

(i) conductors situated on the load side of the point where a reduction occurs in the value of current-carrying capacity, where the conductors are effectively protected against overload by a protective device placed on the supply side of that point,

(ii) conductors which, because of the characteristics of the load, are not likely to carry overload current,

(iii) circuits supplying equipment where unexpected opening of the circuit could cause a greater danger than an overload condition, for example, supply circuits of lifting magnets, exciter circuits of rotating machines,

(iv) secondary circuits of current transformers.

NOTE – The omission of overload protection is recommended for the circuits described in Item (iii) above, but in such cases the provision of an overload alarm should be considered.

Overload protective devices in IT systems

473–4 The provisions of Regulations 473–2 and 473–3 are applicable to installations forming part of an IT system, only where the conductors concerned are protected by a residual current protective device, or all the equipment supplied by the circuit concerned (including the conductors) complies with the protective measure described in Regulations 413–18 to 413–26.

Protection against short circuit

Position of devices for short circuit protection

473–5 A device for protection against short circuit shall be placed at the point where a reduction occurs in the value of current-carrying capacity of the conductors of the installation. This requirement does not apply where the arrangements mentioned in Regulation 473–6 or 473–7 are adopted, and no short-circuit protective device need be provided where Regulation 473–8 applies.

NOTE – A reduction in the value of current-carrying capacity may be caused by a change in cross-sectional area, method of installation, type of cable or conductor, or in environmental conditions.

473–6 The short-circuit protective device may be placed at a point on the load side of that specified in Regulation 473–5 under the following conditions: between the point where the value of current-carrying capacity is reduced and the position of the protective device, the conductors shall –

(i) not exceed 3m in length, and

(ii) be erected in such a manner as to reduce the risk of short circuit to a minimum, and

(iii) be erected in such a manner as to reduce the risk of fire or danger to persons to a minimum.

NOTES: 1 – The condition specified in Item (ii) above may be fulfilled for example, by reinforcing the protection of the conductors against external influences.

2 – The provisions of Regulation 473–6 cannot be applied to any part of an installation in respect of which the short circuit protective device is intended also to afford protection against indirect contact.

473–7 The short circuit protective device may be placed at a point other than that specified in Regulation 473–5, where a protective device on the supply side of that point possesses an operating characteristic such that it protects against short circuit, in accordance with Regulation 434–6, the conductors on the load side of that point.

Omission of devices for short circuit protection

473–8 Devices for protection against short circuit need not be provided for –

(i) conductors connecting generators, transformers, rectifiers, or batteries with their control panels, where short circuit protective devices are placed on those panels,

(ii) certain measuring circuits,

(iii) circuits where disconnection could cause danger in the operation of the installation concerned, such as those mentioned in Regulations 473–3 (iii) and (iv)

provided that the conductors thus not protected against short circuit comply with the conditions specified in Items (ii) and (iii) of Regulation 473–6.

Protection according to the nature of circuits and distribution system

Phase conductors

473–9 Means of detection of overcurrent shall be provided for each phase conductor, and shall cause the disconnection of the conductor in which the overcurrent is detected, but not necessarily the disconnection of other live conductors. Where the disconnection of one phase could cause danger, for example in the supply to three-phase motors, appropriate precautions shall be taken.

Neutral conductor – TN or TT systems

473–10 In TN or TT systems, where the cross-sectional area of the neutral conductor is at least equal or equivalent to that of the phase conductors, it is not necessary to provide overcurrent detection for the neutral conductor or a disconnecting device for this conductor.

473–11 In TN or TT systems, where the cross-sectional area of the neutral conductor is less than that of the phase conductors, overcurrent detection for the neutral conductor shall be provided unless both the following conditions are satisfied:

(i) The neutral conductor is protected against short circuit by the protective device for the phase conductor of the circuit.

(ii) The load is shared as evenly as possible between the various phases of the circuit.

473–12 Where either or both of the conditions specified in Regulation 473–11 are not met, overcurrent detection shall be provided for the neutral conductor, appropriate to the cross-sectional area of that conductor, and the means of detection shall cause the disconnection of the phase conductors but not necessarily of the neutral conductor.

NOTE – The cross-sectional area of the neutral conductor should in any event comply with Regulation 522–9.

Neutral conductor – IT systems

473–13 In IT systems, the distribution of the neutral conductor shall be avoided wherever practicable. Where distribution of the neutral conductor is unavoidable, means of detection of overcurrent shall be provided for the neutral conductor of every circuit, which shall cause disconnection of all the live conductors of the circuit concerned including the neutral conductor. This requirement does not apply where the arrangements described in Regulation 473–14 are adopted.

473–14 In IT systems where the neutral is distributed, means of detection of overcurrent for the neutral conductor need not be provided if either of the following conditions is satisfied:

(i) the neutral conductor concerned is effectively protected against short circuit by a protective device placed on the supply side, for example at the origin of the installation, in accordance with the requirements of Regulation 434–4, or

(ii) the circuit concerned is protected by a residual current device having a rated residual operating current not exceeding 0.15 times the current-carrying capacity of the neutral conductor concerned, and the device is arranged to disconnect all the live conductors of the circuit concerned including the neutral conductor.

474 *Reserved for future use*

475 *Reserved for future use*

476 ISOLATION AND SWITCHING

General

476–1 Every installation shall be provided with means of isolation complying with Section 461. In addition, means of electrical switching off for mechanical maintenance, or means of emergency switching, or both, shall be provided for any parts of the installation to which the relevant section applies.

Where more than one of these functions are to be performed by a common device, the arrangement and characteristics of the device shall satisfy all the requirements of these Regulations for the various functions concerned. Devices for functional switching may serve also for isolation, switching off for mechanical maintenance or emergency switching where they satisfy the relevant requirements.

NOTE – Regulations for selection and erection of devices for isolation and switching are contained in Section 537.

Isolation

476–2 Means of isolation complying with Regulations 461–1 to 461–5 shall be provided at a point as near as practicable to the origin of every installation, without the intervention of any other equipment on which work might need to be done.

476–3 Where an isolator is to be used in conjunction with a circuit breaker as a means of isolating main switchgear for maintenance, it shall be interlocked with the circuit breaker; alternatively, it shall be so placed and/or guarded that it can be operated only by skilled persons.

476–4 Where isolating devices for particular circuits are to be placed remotely from the equipment to be isolated, one or both of the following arrangements shall be made:

(i) The primary means of isolation shall be so placed or guarded that it can be operated only by skilled persons and provision shall be made so that the primary means of isolation cannot be inadvertently returned to the ON position. Where this provision takes the form of a lock or removable handle, the key or handle shall be non-interchangeable with any others used for a similar purpose within the installation.

(ii) An additional isolating device shall be provided adjacent to the equipment.

476–5 Every motor circuit shall be provided with an isolating device or devices which shall disconnect the motor and all equipment, including any automatic circuit breaker, used therewith.

476–6 For electric discharge lighting installations using an open-circuit voltage exceeding low voltage, one or more of the following means shall be provided for the isolation of every self-contained luminaire, or alternatively of every circuit supplying luminaires at a voltage exceeding low voltage:

(i) an interlock on a self-contained luminaire, so arranged that before access can be had to live parts the supply is automatically disconnected, such means being additional to the switch normally used for controlling the circuit,

(ii) effective local means for the isolation of the circuit from the supply, such means being additional to the switch normally used for controlling the circuit,

(iii) a switch having a lock or removable handle, or a distribution board which can be locked, in either case complying with Regulation 476–4 (i).

Switching off for mechanical maintenance

476–7 A means of switching off for mechanical maintenance shall be provided for every circuit supplying an electric motor, or equipment having electrically heated surfaces which can be touched, or electromagnetic equipment for operations from which mechanical accidents could arise.

476–8 Where a switch mounted on an appliance or luminaire is intended to serve as a means of switching off for mechanical maintenance, the connections shall be so arranged that the appliance or luminaire can be dismantled to the extent necessary for mechanical maintenance without thereby exposing any parts which remain live when the switch is open. Any conductors or cables which then remain live shall be as short as possible, and separated from any other live conductors or cables by screens or earthed metal or suitable barriers of insulating material.

Emergency switching

476–9 For every emergency switching device, account shall be taken of the intended use of the premises so that access to the device is not likely to be impeded in the conditions of emergency foreseen.

476–10 Where greater danger would arise from incorrect operation of emergency switching (as for example by inadvertent disconnection of safety services), the means of emergency switching may be arranged so as to be suitable for operation by skilled persons or instructed persons only.

476–11 Means of emergency switching shall be provided in every place where a machine driven by electric means may give rise to danger, and shall be readily accessible and easily operated by the person in charge of the machine. Where more than one means of manually stopping the machine is provided and danger might be caused by unexpected restarting, means shall be provided to prevent such restarting.

476–12 A fireman's emergency switch shall be provided for –

– exterior discharge lighting installations operating at a voltage exceeding low voltage, and

– interior discharge lighting installations operating unattended at a voltage exceeding low voltage.

For the purpose of this regulation, an installation in a closed market or in an arcade is considered to be an exterior installation. A temporary installation in a permanent building used for exhibitions is considered not to be an exterior installation. This requirement does not apply to a portable discharge lighting luminaire or sign of rating not exceeding 100W and fed from a readily accessible socket outlet.

476–13 Every fireman's emergency switch provided for compliance with Regulation 476–12 shall comply with all the relevant requirements of the following items (i) to (iv):

(i) For exterior installations, the switch shall be outside the building and adjacent to the discharge lamp(s), or alternatively a notice indicating the position of the switch shall be placed adjacent to the discharge lamp(s) and a nameplate shall be fixed near the switch so as to render it clearly distinguishable.

(ii) For interior installations, the switch shall be in the main entrance to the building or in another position to be agreed with the local fire authority.

(iii) The switch shall be placed in a conspicuous position, reasonably accessible to firemen and, except where otherwise agreed with the local fire authority, at not more than 2.75m from the ground.

(iv) Where more than one switch is installed on any one building, each switch shall be clearly marked to indicate the installation or part of the installation which it controls, and the local fire authority shall be notified accordingly.

NOTE – Wherever practicable, all exterior installations on any one building should be controlled by a single fireman's switch. Similarly, all internal installations in any one building should be controlled by a single fireman's switch independent of the switch for any external installation.

Other requirements for switching

476–14 In situations where the requirements of Section 463 for emergency switching are not applicable, the requirements of the following Regulations 476–15 to 476–20 shall be satisfied, either –

– by the means of isolation and/or the means of switching off for mechanical maintenance provided for compliance with Sections 461 and 462, or

– by the arrangements for switching of equipment for its normal service (functional switching), or

– by the provision of suitable additional means of switching.

476–15 A main switch or circuit breaker shall be provided for every installation which shall interrupt all live conductors of the installation, provided that for a 4-wire three-phase a.c. supply the linked switch or linked circuit breaker may be arranged to disconnect the phase conductors only and a link may be inserted in the neutral conductor; such a link shall be arranged so that it is in contact before the linked switch can be closed, or shall be securely fixed by bolts or screws.

476–16 Every circuit and final circuit shall be provided with means of interrupting the supply on load and in any fault conditions foreseen. A group of circuits may be switched by a common device. Additionally, such means shall be provided for every circuit or other part of the installation which it may be necessary for safety reasons to switch independently of other circuits or other parts of the installation. This regulation does not apply to short connections between the origin of the installation and the consumer's main switchgear.

476–17 Every appliance or luminaire connected to the supply other than by means of a plug and socket outlet, shall be controlled by a switch or switches separate from the appliance or luminaire and in a readily accessible position, subject to the provisions of Regulations 476–18 or 476–19 where applicable. For an appliance fitted with heating elements which can be touched, the switch shall be a linked switch arranged to break all the circuit conductors including the neutral. For the purpose of this regulation the sheath of a silica glass sheathed element is regarded as part of the element.

476–18 An appliance or luminaire may be controlled by a switch mounted on the appliance or luminaire, subject to Regulation 476–8 where the switch is intended to serve as a means of switching off for mechanical maintenance.

476–19 The switch or switches providing control of comprehensive heating or lighting installations comprising more than one appliance or luminaire, may be installed in a separate room.

476–20 Every fixed or stationary household cooking appliance shall be controlled by a switch separate from the appliance and placed within 2m of the appliance. Where two stationary cooking appliances are installed in one room of household premises, one switch may be used to control both appliances provided that neither appliance is more than 2m from the switch.

PART 5

Selection and erection of equipment

CONTENTS

CHAPTER 51 **COMMON RULES**

510 General

511 Compliance with standards

512 Operational conditions and external influences

513 Accessibility

514 Identification and notices

515 Mutual detrimental influence

CHAPTER 52 **CABLES, CONDUCTORS AND WIRING MATERIALS**

521 Selection of types of wiring systems

522 Operational conditions

523 Environmental conditions

524 Identification

525 Prevention of mutual detrimental influence

526 Accessibility

527 Joints and terminations

528 Fire barriers

529 Supports, bends and space factors

CHAPTER 53 **SWITCHGEAR** (for protection, isolation and switching)

530 Common requirements

531 Devices for protection against electric shock

532 *(Reserved for future use)*

533 Overcurrent protective devices

534 *(Reserved for future use)*

535 *(Reserved for future use)*

536 *(Reserved for future use)*

537 Isolating and switching devices

CHAPTER 54 **EARTHING ARRANGEMENTS AND PROTECTIVE CONDUCTORS**

541 General

542 Connections to Earth

543 Protective conductors

544 Earthing, and protective conductors, for fault-voltage operated protective devices

545 *(Reserved for future use)*

546 Combined protective and neutral (PEN) conductors

547 Protective bonding conductors

CHAPTER 55 **OTHER EQUIPMENT**

551 Transformers

552 Rotating machines

553 Accessories

554 Current-using equipment

CHAPTER 51

COMMON RULES

CONTENTS

PART 5

Selection and erection of equipment

CHAPTER 51

COMMON RULES

510 GENERAL

510–1 Every item of equipment shall be selected and erected so as to comply with the requirements stated in the following regulations of this chapter and the relevant regulations in other chapters of these Regulations.

511 COMPLIANCE WITH STANDARDS

511–1 Every item of equipment shall comply with the relevant requirements of the current edition of the applicable British Standard.

NOTES: 1 – The current edition is that edition, with amendments if any, pertaining at a date to be agreed by the parties to the contract concerned.

2 – Where equipment manufactured in compliance with the current edition of the applicable British Standard is used, it is assumed that the construction of the equipment satisfies the requirements of these Regulations, provided that the intended scope of application of the British Standard is appropriate to the use intended.

3 – Where the particular equipment to be used is not covered by a British Standard, the designer or other person responsible for specifying the installation should verify also that the equipment provides the same degree of safety as that afforded by compliance with these Regulations. (See also the last two paragraphs of the Preface to these Regulations concerning standards other than British Standards, and Regulation 12–6 concerning the ANT Scheme).

4 – For some equipment within the scope of these Regulations, certification schemes are operated by the British Standards Institution, the British Electrotechnical Approvals Board, the British Approvals Service for Electric Cables and the Association of Short Circuit Testing Authorities.

512 OPERATIONAL CONDITIONS AND EXTERNAL INFLUENCES

Voltage

512–1 All equipment shall be suitable for the nominal voltage (r.m.s. value for a.c.) of the installation or the part of the installation concerned.

In an IT system, if the neutral conductor is distributed, equipment connected between phase and neutral shall be insulated for the nominal voltage between phases.

NOTES: 1 – Where inductive or capacitive equipment is concerned, the design of switches and circuit breakers should be adequate for such service.

2 – For certain equipment, it may be necessary to take account of the highest and/or lowest voltage likely to occur in normal service.

Current

512–2 All equipment shall be suitable for:

(i) the design current (r.m.s. value for a.c.),

(ii) the current likely to flow in abnormal conditions for such periods of time as are determined by the characteristics of the protective devices concerned.

NOTE – Where inductive or capacitive equipment is concerned, the design of switches and circuit breakers should be adequate for such service.

Frequency

512–3 If frequency has an influence on the characteristics of the equipment, the rated frequency of the equipment shall correspond to the frequency of the current in the circuit concerned.

Power

512–4 Equipment to be selected on the basis of its power characteristics shall be suitable for the duty demanded of the equipment, taking account of the load factor and the normal operational conditions.

Compatibility

512–5 All equipment shall be selected and erected so that it will not cause harmful effects on other equipment or impair the supply during normal service including switching operations.

External influences

512–6 All equipment shall be of a design appropriate to the situation in which it is to be used and its mode of installation shall take account of the conditions likely to be encountered.

NOTE – See also the Note to Chapter 32.

513 ACCESSIBILITY

513–1 All equipment shall be arranged so as to facilitate its operation, inspection and maintenance and access to its connections. Such facilities shall not be significantly impaired by mounting equipment in enclosures or compartments. This regulation does not apply to joints in cables where Section 526 allows such joints to be inaccessible.

514 IDENTIFICATION AND NOTICES

General

514–1 Labels or other suitable means of identification shall be provided to indicate the purpose of switchgear and controlgear, unless there is no possibility of confusion.

Where the operation of switchgear and controlgear cannot be observed by the operator and where this might cause a danger, a suitable indicator, complying with BS 4099 where applicable, shall be fixed in a position visible to the operator.

NOTE – See also Chapter 52 for identification of conductors and cores of cables, and Section 554 for requirements for notices in discharge lighting installations operating at a voltage exceeding low voltage.

Protective devices

514–2 Protective devices shall be arranged and identified so that the circuits protected may be easily recognised; for this purpose it may be convenient to group them in distribution boards.

Diagrams

514–3 Diagrams, charts or tables shall be provided indicating in particular –

– the type and composition of circuits (points of utilisation served, number and size of conductors, type of wiring), and

– the information necessary for the identification of the devices performing the functions of protection, isolation and switching, and their locations, and

– a description of the method used for compliance with Regulation 413–3.

For simple installations the foregoing information may be given in a schedule.

Any symbols used shall comply with BS 3939.

Warning notices – voltage

514–4 Every item of equipment or enclosure within which a voltage exceeding 250 volts exists, and where the presence of such a voltage would not normally be expected, shall be so arranged that before access is gained to live parts, a warning of the maximum voltage present is clearly visible.

Where terminals or other fixed live parts between which a voltage exceeding 250 volts exists are housed in separate enclosures or items of equipment which, although separated, can be reached simultaneously by a person, a notice shall be placed in such a position that anyone gaining access to live parts is warned of the maximum voltage which exists between those parts.

Means of access to all live parts of switchgear and other fixed live parts where different nominal voltages exist shall be marked to indicate the voltages present.

Notice – periodic inspection and testing

514–5 A notice of such durable material as to be likely to remain easily legible throughout the life of the installation, shall be fixed in a prominent position at or near the main distribution board of every installation upon completion of the work. The notice shall be inscribed in indelible characters not smaller than those here illustrated (11-point) and shall read as follows:

'IMPORTANT

This installation should be periodically inspected and tested, and a report on its condition obtained, as prescribed in the Regulations for Electrical Installations issued by The Institution of Electrical Engineers.

Date of last inspection ....

Recommended date of next inspection ....'

NOTE – For the recommended number of years intervening between periodic inspections see the footnote to the Inspection Certificate in Appendix 16.

Warning notice – caravans

514–6 In all touring (mobile) caravans, a notice of durable material shall be fixed near the main switch inside the caravan, bearing in indelible and easily legible characters the text shown below:

INSTRUCTIONS FOR ELECTRICITY SUPPLY

On arrival at caravan site

1. Before connecting the caravan installation to the mains supply, check that

   (a) the mains supply is suitable for your installation and appliances, i.e. whether it is a.c. or d.c. and whether it is at the correct voltage and frequency, and

   (b) your installation will be properly earthed. Never accept a supply from a socket outlet or plug having only two pins, or from a lighting outlet.

   (c) any residual current device (earth leakage circuit breaker) in the mains supply to the caravan has been tested within the last month.

   In case of doubt, consult the site owner or his agent.

2. **MAKE SURE THAT THE SWITCH AT THE SITE SUPPLY POINT IS OFF.**

3. Remove any cover from the electricity inlet provided on the caravan, and insert the connector of the supply flexible cable obtained from the site owner.

4. Remove any cover from the socket outlet provided at the site supply point, and connect the plug

at the other end of the supply flexible cable to this. Switch on the main switch at the site supply point.

IN CASE OF DIFFICULTY CONSULT AN APPROVED ELECTRICAL INSTALLATION CONTRACTOR (WHO MAY BE THE LOCAL ELECTRICITY BOARD). IT IS DANGEROUS TO ATTEMPT MODIFICATIONS AND ADDITIONS YOURSELF. LAMPHOLDER-PLUGS (BAYONET-CAP ADAPTORS) SHOULD NOT IN ANY CIRCUMSTANCES BE USED.

On leaving caravan site

5. Reverse the procedure described in Paragraphs 3 and 4 above.

IT IS IMPORTANT THAT THE MAIN SWITCH AT THE SITE SUPPLY POINT SHOULD BE SWITCHED OFF, THE SUPPLY FLEXIBLE CABLE DISCONNECTED, AND ANY COVER REPLACED ON THE SOCKET OUTLET AT THE SITE SUPPLY POINT. IT IS DANGEROUS TO LEAVE THE SUPPLY SOCKET OR SUPPLY FLEXIBLE CABLE LIVE.

Periodically

6. Preferably not less than once a year, the caravan electrical installation should be inspected and tested and a report on its condition obtained as prescribed in the Regulations for Electrical Installations, published by the Institution of Electrical Engineers.

Warning notices — earthing and bonding connections

514–7 A permanent label durably marked with the words 'Safety electrical connection – do not remove', in legible type not less than 4.75mm high, shall be permanently fixed in a visible position at or near –

– the point of connection of every earthing conductor to an earth electrode, and

– the point of connection of every bonding conductor to extraneous conductive parts.

Notice for socket outlets

514–8 A notice shall be provided on or near every socket outlet specifically intended to supply equipment to be used outside the zone created by the main equipotential bonding of the installation (see Regulation 471–12). The notice shall be of such durable material as to be likely to remain easily legible throughout the life of the installation and shall read as follows: 'FOR EQUIPMENT OUTDOORS'.

515 MUTUAL DETRIMENTAL INFLUENCE

515–1 All electrical equipment shall be selected and erected so as to avoid any harmful influence between the electrical installation and any non-electrical installations envisaged.

515–2 Where equipment carrying currents of different types or at different voltages is grouped in a common assembly, all equipment using any one type of current or any one voltage shall, wherever necessary, be effectively segregated from equipment of any other type, to avoid mutual detrimental influence.

CHAPTER 52

CABLES, CONDUCTORS, AND WIRING MATERIALS

CONTENTS

521	**SELECTION OF TYPES OF WIRING SYSTEMS**
521–1 to **521–4**	Non-flexible cables and conductors for low voltage
521–5 and **521–6**	Flexible cables and flexible cords for low voltage
521–7	Cables for extra-low voltage
521–8	Cables for a.c. circuits — electromagnetic effects
521–9 to **521–11**	Conduits and conduit fittings
521–12	Trunking, ducting and fittings
521–13 and **521–14**	Methods of installation of cables and conductors
522	**OPERATIONAL CONDITIONS**
522–1 to **522–7**	Current-carrying capacity
522–8	Voltage drop
522–9	Minimum cross-sectional area of neutral conductors
522–10	Electromechanical stresses
523	**ENVIRONMENTAL CONDITIONS**
523–1 to **523–6**	Ambient temperature
523–7 to **523–15**	Presence of water or moisture
523–16	Dust
523–17 and **523–18**	Corrosive or polluting substances
523–19 to **523–33**	Mechanical stresses
523–34	Damage by fauna
523–35	Solar radiation
524	**IDENTIFICATION**
524–1 and **524–2**	General
524–3	Non-flexible cables and conductors
524–4 and **524–5**	Flexible cables and flexible cords
525	**PREVENTION OF MUTUAL DETRIMENTAL INFLUENCE**
525–1 to **525–9**	Between low voltage circuits and circuits of other categories
525–10 to **525–12**	Between electrical services and exposed metalwork of other services

CHAPTER 52

CABLES, CONDUCTORS, AND WIRING MATERIALS

NOTE – The requirements of this chapter do not apply to heating cables and conductors, which in these Regulations are treated as current-using equipment, see Section 554. For earthed-concentric wiring systems, see Chapter 54. For cables for discharge lighting circuits operating at a voltage exceeding low voltage, see Section 554.

521 SELECTION OF TYPES OF WIRING SYSTEMS

Non-flexible cables and conductors for low voltage

521–1 Every non-flexible cable operating at low voltage shall be selected from one of the following types and shall comply with the appropriate British Standard referred to below so far as this is applicable. In cables of every type, conductors of cross-sectional area 10mm^2 or less shall be of copper or copperclad aluminium. Any of the types of cable sheathed with p.v.c. or lead, or having a h.o.f.r. sheath or an oil-resisting and flame-retardant sheath, if intended for aerial suspension, may incorporate a catenary wire.

(i) Non-armoured p.v.c.-insulated cables (BS 6004, BS 6231 Type B, or BS 6346).

(ii) Armoured p.v.c.-insulated cables (BS 6346).

(iii) Split-concentric copper-conductor p.v.c.-insulated cables (BS 4553).

(iv) Rubber insulated cables (BS 6007).

(v) Impregnated-paper-insulated cables, lead-sheathed (BS 6480).

(vi) Armoured cables with thermosetting insulation (BS 5467).

(vii) Mineral-insulated cables (BS 6207, Part 1 or Part 2), with, where appropriate, fittings to BS 6081.

(viii) Consac cables (BS 5593).

(ix) Cables approved under Regulation 12 of the Electricity Supply Regulations.

NOTES: 1 – See Appendix 10 concerning selection of types of cables for particular uses.

2 – For information on the cables mentioned in item (ix) reference should be made to the electricity supply undertaking.

3 – Mineral-insulated cables should not be used in discharge lighting circuits unless suitable precautions are taken to avoid excessive voltage surges.

521–2 Paper-insulated cables shall be of a type which will comply with the requirements of BS 6480 for non-draining cables, where drainage of the impregnating compound would otherwise be liable to occur.

521–3 Busbars and busbar connections shall comply with BS 5486, Part 1.

521–4 Every conductor, other than a cable, for use as an overhead line operating at low voltage shall be selected from one of the following types and shall comply with the appropriate British Standard referred to below.

(i) Hard-drawn copper or cadmium-copper conductors (BS 125).

(ii) Hard-drawn aluminium and steel-reinforced aluminium conductors (BS 215).

(iii) Aluminium-alloy conductors (BS 3242).

(iv) Conductors covered with p.v.c. for overhead power lines (BS 6485 Type 8).

Flexible cables and flexible cords for low voltage

521–5 Every flexible cable and flexible cord for use at low voltage shall be selected from one of the following types and shall comply with the appropriate British Standard referred to below so far as this is applicable. This regulation does not apply to a flexible cord forming part of a portable appliance or luminaire where the appliance or luminaire as a whole is the subject of, and complies with, a British Standard or to special flexible cables and flexible cords for combined power and telecommunication wiring (see Regulation 525–8).

(i) Insulated flexible cords (BS 6500).

(ii) Rubber-insulated flexible cables (BS 6007).

(iii) PVC-insulated flexible cables (non-armoured) (BS 6004).

(iv) Braided travelling cables for lifts (BS 6977).

(v) Rubber-insulated flexible trailing cables for quarries and miscellaneous mines (BS 6116).

Such insulated flexible cables and flexible cords may incorporate a flexible armour of galvanised steel or phosphor-bronze, or a screen of tinned copper-wire braid.

521–6 Every flexible cable and flexible cord shall be selected from one of the following types:

(i) Braided circular.

(ii) Unkinkable.

(iii) Circular sheathed.

(iv) Flat twin sheathed.

(v) Parallel twin, only for the wiring of luminaires (where permitted by BS 4533).

(vi) Twisted twin non-sheathed, only for the wiring of luminaires (where permitted by BS 4533).

(vii) Braided circular twin and three-core, insulated with glass fibre, where permitted by Regulation 523–28.

(viii) Single-core p.v.c.-insulated non-sheathed flexible cables complying with BS 6004 installed in accordance with Regulation 523–29.

NOTES: 1 – The types of flexible cord for use with particular types of appliance and luminaire are specified in BS 3456 and BS 4533 respectively. See also the Electrical Equipment (Safety) Regulations 1975*, and the Electrical Equipment (Safety) (Amendment) Regulations 1976*.

2 – See Appendix 10 concerning selection of types of flexible cables and flexible cords for particular uses.

Cables for extra-low voltage

521–7 Cables for use at extra-low voltage shall have adequate insulation, and further protection if necessary, so as to prevent danger.

Cables for a.c. circuits – electromagnetic effects

521–8 Single-core cables armoured with steel wire or tape shall not be used for a.c. Conductors of a.c. circuits installed in ferrous enclosures shall be arranged so that the conductors of all phases and the neutral conductor (if any) are contained in the same enclosure. Where such conductors enter ferrous enclosures they shall be arranged so that the conductors are not separated by a ferrous material or provisions shall be made to prevent circulating eddy currents.

Conduits and conduit fittings

521–9 Conduits and conduit fittings shall comply with the appropriate British Standard referred to below.

(i) Steel conduit and fittings (BS 31).

(ii) Flexible steel conduit (BS 731).

(iii) Steel conduit and fittings with metric threads (BS 4568).

(iv) Non-metallic conduits and fittings (BS 4607).

*H.M. Stationery Office

521–10 In conduit systems, the conduits for each circuit shall be completely erected before any cable is drawn in. This requirement does not apply to prefabricated conduit systems which are not wired in situ.

521–11 In the prefabrication of conduit systems which are not to be wired in situ, adequate allowance shall be made for variations in building dimensions so that the conduits or cables are not subjected to tension or other mechanical strain during installation. Adequate precautions shall also be taken to prevent damage to such systems during installation and any subsequent building operations, especially against deformation of the conduits and damage to any exposed cable ends.

Trunking, ducting and fittings

521–12 Trunking, ducting and fittings shall comply, where applicable, with BS 4678 or be of insulating material having the ignitability characteristic 'P' as specified in BS 476, Part 5.

Methods of installation of cables and conductors

521–13 Methods of installation of cables and conductors in common use for which these Regulations specifically provide are described in Appendix 9. The use of other methods is not precluded where specified by a suitably qualified electrical engineer, provided that the applicable requirements of this chapter are complied with.

521–14 Ducts cast in situ in concrete, by means of a suitable form laid before the concrete is poured, into which cables are to be drawn (whether or not formers are retained in position after the concrete has set) shall be so formed that the radial thickness of concrete or screed surrounding the cross-section of the completed duct is not less than 15mm at every point.

522 OPERATIONAL CONDITIONS

Current-carrying capacity

522–1 The cross-sectional area of every cable conductor shall be such that its current-carrying capacity is not less than the maximum sustained current which will normally flow through it. For the purposes of this regulation the limiting temperature to which the current-carrying capacity relates shall not exceed that appropriate to the type of cable insulation concerned. This regulation does not apply to conductors on switchboards complying with Regulation 522–2.

NOTE – Appendix 9 gives limiting temperatures on which the current-carrying capacities of various types of cable are based, together with the method for determining the cross-sectional area of a cable conductor for compliance with Regulation 522–1.

522–2 Busbars, busbar connections and bare conductors forming part of the equipment of switchboards shall comply as regards current-carrying capacity and limits of temperature with the requirements of BS 5486, Part 1.

522–3 Cables connected in parallel shall be of the same type, cross-sectional area, length and disposition and be arranged so as to carry substantially equal currents.

522–4 In determining the current-carrying capacity of bare conductors, account shall be taken of the arrangements made for their expansion and contraction, their joints, and the physical limitations of the metal of which they are made.

NOTE – It is recommended that the maximum operating temperature of bare conductors should not exceed 90°C.

522–5 Where cables are to be connected to bare conductors or busbars it shall be verified that their type of insulation and/or sheath is suitable for the maximum operating temperature of the bare conductors or busbars. Alternatively the insulation and/or sheath of the cables shall be removed for a distance of 150mm from the connection and replaced if necessary by suitable heat-resisting insulation.

522–6 Where a cable is to be run for a significant length in a space to which thermal insulation is likely to be applied, the cable shall wherever practicable be fixed in a position such that it will not be covered by the thermal insulation. Where fixing in such a position is impracticable, the current-carrying capacity of the cable shall be appropriately reduced.

NOTE – For a cable installed in a thermally insulating wall or above a thermally insulated ceiling, the cable being in contact with a thermally conductive surface on one side, the rating factor to be applied may, in the absence of more precise information, be taken as 0.75 times the current-carrying capacity for that cable clipped direct to a surface and open. For a cable likely to be totally surrounded by thermally insulating material, the applicable rating factor may be as low as 0.5.

522–7 Metallic sheaths and/or non-magnetic armour of all single-core cables in the same circuit shall normally be bonded together at both ends of their run (solid bonding). Alternatively, where specified by a suitably qualified electrical engineer, such cables having conductors of cross-sectional area exceeding 50mm² may be bonded together at one point in their run (single point bonding) with suitable insulation at the open-circuit end, in which case the length of the cables from the bonding point shall be limited so that voltages from sheaths and/or armour to Earth do not –

– exceed 25 volts and do not cause corrosion when the cables are carrying their full load current, and

– do not cause danger or damage to property when the cables are carrying short circuit current.

NOTE – The relevant current-carrying capacities indicated in Appendix 9 relate to solid bonding only. The use of single-point bonding permits a higher current-carrying capacity.

Voltage drop

522–8 The size of every bare conductor or cable conductor shall be such that the drop in voltage from the origin of the installation to any point in that installation does not exceed 2.5% of the nominal voltage when the conductors are carrying the full load current, but disregarding starting conditions. This requirement does not apply to wiring fed from an extra-low voltage source.

NOTES: 1 – Values of voltage drop per ampere per metre run of cable(s) are given in the Tables of Appendix 9.

2 – In some instances a cable which gives a voltage drop less than that permissible under Regulation 522–8 may be necessary for satisfactory starting of motors. Account should be taken of the effects of motor starting currents on other equipment.

Where an allowance is made for diversity in accordance with Section 311, this may be taken into account in calculating voltage drop.

Minimum cross-sectional area of neutral conductors

522–9 For polyphase circuits in which imbalance may occur in normal service, through significant inequality of loading or of power factors in the various phases, or through the presence of significant harmonic currents in the various phases, the neutral conductor shall have a cross-sectional area adequate to afford compliance with Regulation 522–1 for the maximum current likely to flow in it. (See also Section 473 as to overcurrent protection of neutral conductors).

For polyphase circuits in which serious imbalance is unlikely to be sustained in normal service, other than discharge lighting circuits, the use of multicore cables incorporating a reduced neutral conductor in accordance with the appropriate British Standard is recognised. Where single-core cables are used in such circuits, the neutral conductor may have a reduced cross-sectional area appropriate to the expected value of the neutral current.

In any circuit where the load is predominantly due to discharge lighting, the neutral conductor shall have a cross-sectional area not less than that of the phase conductor(s).

Electromechanical stresses

522–10 All conductors and cables shall have adequate strength and be so installed as to withstand the electromechanical forces that may be caused by any current they may have to carry in service, including short circuit current.

523 ENVIRONMENTAL CONDITIONS

Ambient temperature

523–1 The type and current-carrying capacity of every conductor, cable and flexible cord, termination and joint shall be selected so as to be suitable for the highest operating temperature likely to occur in normal service, account being taken of any transfer of heat from any accessory, appliance or luminaire to which the conductor, cable or flexible cord is connected.

523–2 For cables, other than heating cables, installed in a heated floor or other heated part of a building, the maximum normal operating temperature of that part of the floor etc., in which the cable is installed shall be taken as the relevant ambient temperature.

NOTE – For guidance on temperatures of floors in off-peak floor warming systems, see CP 1018.

523–3 Parts of a cable or flexible cord within an accessory, appliance or luminaire shall be suitable for the temperatures likely to be encountered, as determined in accordance with Regulation 523–1, or shall be provided with additional insulation suitable for those temperatures. Such additional insulation shall be fitted over the individual cores of the cable or flexible cord in such a way that the normal insulation of the cores is not relied upon to prevent a short circuit between conductors or an earth fault.

NOTE – Exposure of plastics-insulated cables to high temperature, even for short periods, may cause the insulation to soften. Continuous exposure of p.v.c. compounds to temperatures above 115°C may contribute to the formation of corrosive products which can attack conductors and other metalwork.

523–4 In determining the normal operational conditions of conductors and cables, account need not be taken of the minimum ambient temperature likely to occur. However, precautions shall be taken to avoid risk of mechanical damage to cables susceptible to low temperatures.

523–5 The enclosures of wiring systems for conductors and cables shall be selected and installed so that they are suitable for the extremes of ambient temperature to which they are likely to be exposed in normal service. If a non-metallic or composite outlet box is used for the suspension of, or is in contact with, a luminaire and where a thermoplastic material (e.g. p.v.c.) is the principal load-bearing member, care shall be taken that the temperature of the box does not exceed 60°C and that the mass suspended from the box does not exceed 3 kg.

NOTE – In determining the ambient temperature of such enclosures, account should be taken of the maximum normal operating temperatures of the conductors of cables installed within them.

523–6 In every vertical channel, duct, ducting or trunking containing conductors or cables, internal barriers shall be provided so as to prevent the air at the top of the channel, duct, ducting or trunking from attaining an excessively high temperature. The distance between barriers shall be the distance between floors or 5m whichever is the less.

NOTE – The fire barriers specified in Regulation 528–1 may serve also for compliance with Regulation 523–6.

Presence of water or moisture

523–7 Every wiring system shall either be installed where it will not be exposed to rain, dripping water, steam, condensed water or accumulations of water or be of a type designed to withstand such exposure.

523–8 In damp situations and wherever they are exposed to the weather, all metal sheaths and armour of cables, metal conduit, ducts, ducting, trunking, clips and their fixings, shall be of corrosion-resisting material or finish and shall not be placed in contact with other metals with which they are liable to set up electrolytic action.

523–9 Copperclad aluminium conductors shall not be used in situations where the terminations of the conductors are likely to be exposed to sustained wet conditions.

NOTE – Regulations 523–8 and 523–9 do not apply to situations which may be only initially damp during building construction.

523–10 A plain aluminium conductor shall not be placed in contact with a terminal of brass or other metal having a high copper content, unless the terminal is suitably plated or other precautions are taken to maintain electrical continuity.

523–11 In any situation, the exposed conductor and insulation at terminations and joints of cables insulated with impregnated paper shall be protected from ingress of moisture by being suitably sealed.

523–12 The ends of mineral-insulated cables shall be protected from moisture by being suitably sealed and the insulation shall be thoroughly dry before the sealing material is applied. Such sealing material, and any material used to insulate the conductors where they emerge from the insulation, shall have adequate insulating and moisture-proofing properties, and shall retain those properties throughout the range of temperatures to which they may be subjected in service.

523–13 In damp situations, enclosures for cores of sheathed cables from which the sheath has been removed and for non-sheathed cables at terminations of conduit, duct, ducting or trunking systems shall be damp-proof and corrosion-resistant. Every joint in a cable shall be suitably protected against the effects of moisture.

523–14 Conduit systems not designed to be sealed shall be provided with drainage outlets at any points in the installation where moisture might otherwise collect.

523–15 Every entry to finished ducts, ducting or trunking shall be placed so as to prevent the ingress of water, or be protected against such ingress.

Dust

523–16 Enclosures for conductors and their joints and terminations in onerous dust conditions shall have the degree of protection IP 5X (see BS 5490).

Corrosive or polluting substances

523–17 All metalwork of wiring systems shall either be installed where it will not be exposed to corrosive substances, or be of a type or be protected so as to withstand such exposure (see Appendix 10). Non-metallic materials of wiring systems shall not be placed in contact with materials likely to cause chemical deterioration of the wiring systems. Such materials shall either be installed where they will not be exposed to contact with oil, creosote, and similar hydrocarbons, or be of a type designed to withstand such exposure.

523–18 Soldering fluxes which remain acidic or corrosive at the completion of the soldering process shall not be used unless suitable precautions are taken to neutralise their effect.

Mechanical stresses

523–19 All conductors and cables shall be adequately protected against any risk of mechanical damage to which they may be liable in normal conditions of service.

523–20 Where cables are installed under floors or above ceilings they shall be run in such positions that they are not liable to be damaged by contact with the floor or the ceiling or their fixings. Where the cable passes through a timber joist within a floor or ceiling construction (e.g., under floorboards), the cable shall be 50mm measured vertically from the top, or bottom as appropriate, of the joist. Alternatively, cables not protected by an earthed metallic sheath shall be protected by enclosure in earthed steel conduit securely supported, or by equivalent mechanical protection sufficient to prevent penetration of the cable by nails, screws, and the like.

523–21 Where cables pass through holes in metalwork, precautions shall be taken to prevent abrasion of the cables on any sharp edges.

523–22 Non-sheathed cables for fixed wiring shall be enclosed in conduit, duct, ducting or trunking. Such non-sheathed cables shall not be installed in ducts cast in situ in concrete where any part of the completed duct is formed by the concrete or screed.

523–23 Cables buried direct in the ground shall be of a type incorporating an armour or metal sheath or both, or be of the p.v.c.-insulated concentric type. Such cables shall be marked by cable covers or a suitable marking tape and be buried at a sufficient depth to avoid their being damaged by any disturbance of the ground reasonably likely to occur during the normal use of the premises.

523–24 Cables to be installed in underground ducts, conduits or pipes shall be of a type incorporating a sheath and/or armour suitably resistant to any mechanical damage likely to be caused during drawing in.

523–25 Cables to be installed outdoors on walls and the like shall incorporate a sheath and/or armour suitably resistant to any mechanical damage likely to occur, or be contained in a conduit system or other enclosure affording adequate protection against such damage.

523–26 Cables for overhead wiring between a building and a point of utilisation not attached thereto (e.g. another building) shall be so placed and at such a height as to be out of reach of any sources of mechanical damage reasonably to be foreseen in the normal use of the premises. Alternatively, for spans in situations inaccessible to vehicular traffic, such cables may be installed in conduit or other enclosure affording adequate protection against such damage.

NOTES: 1 – Bare or lightly insulated overhead conductors are required to be placed out of reach of persons and livestock, for compliance with Regulation 412–9.

2 – See Appendix 11 for notes on methods of support.

523–27 Flexible cords, where they are exposed to risk of mechanical damage, shall be of a type sheathed with rubber or p.v.c. and where necessary shall also be armoured; provided that for domestic and similar applications where flexible cords are subject only to moderate bending and/or wear, unkinkable flexible cords complying with BS 6500 may be used.

523–28 Braided circular twin and three-core flexible cords insulated with glass fibre shall be used only for luminaires or for other applications where the cord is not subject to abrasion or undue flexing.

523–29 Flexible cords shall not be used as fixed wiring, except as permitted by Regulation 523–31, unless contained in an enclosure affording mechanical protection.

523–30 Flexible cables or flexible cords shall be used for connections to portable equipment. For the purpose of this regulation an electric cooker of rated input exceeding 3kW is considered not to be portable. Such flexible cables or flexible cords shall be of suitable length to avoid undue risk of mechanical damage.

523–31 Exposed lengths of flexible cable or flexible cord used for final connections to fixed equipment shall be as short as possible and connected to the fixed wiring by a suitable accessory or an enclosure complying with Regulation 527–4 or, where Chapters 43 and 46 so require, by a suitable device or devices for overcurrent protection, isolation, and switching.

523–32 Where a flexible cord supports or partly supports a luminaire, the maximum mass supported by the cord shall not exceed the appropriate value indicated below:

Nominal cross-sectional area of conductor (mm^2)	Maximum mass (kg)
0.5	2
0.75	3
1.0	5

523–33 In assessing risks of mechanical damage to cables, account shall be taken of any mechanical strains likely to be imposed during the normal process of erection of the cables.

Damage by fauna

523–34 In premises intended for livestock all fixed wiring systems shall be inaccessible to livestock. Cables liable to attack by vermin shall be of a suitable type or be suitably protected.

Solar radiation

523–35 Cables and wiring systems installed in positions which may be exposed to direct sunlight shall be of a type resistant to damage by ultra-violet light.

524 IDENTIFICATION

General

524–1 The colour combination green and yellow is reserved exclusively for identification of protective conductors and shall not be used for any other purpose. In this combination one of the colours shall cover at least 30% and at most 70% of the surface being coloured, while the other colour covers the remainder of the surface.

524–2 Electrical conduits where required to be distinguished from pipelines of other services shall use orange as the basic identification colour in compliance with BS 1710.

Non-flexible cables and conductors

524–3 Every single-core non-flexible cable and every core of a twin or multicore non-flexible cable for use as fixed wiring shall be identifiable at its terminations and preferably throughout its length, by the appropriate method described in items (i) to (v) below.

(i) For rubber-insulated and p.v.c.-insulated cables, the use of core colours in accordance with the requirements of Table 52A, or the application at terminations of tapes, sleeves or discs of the appropriate colours prescribed in the Table.

(ii) For armoured p.v.c.-insulated auxiliary cables, as an alternative to the method described in Item (i) above, the use of numbered cores in accordance with BS 6346.

(iii) For paper-insulated cables, the use of numbered cores in accordance with BS 6480; provided that the numbers 1, 2 and 3 shall signify phase conductors, the number 0 the neutral conductor, and the number 4 the fifth ('special-purpose') core if any.

(iv) For cables with thermosetting insulation, the use of core colours in accordance with the requirements of Table 52A, or alternatively the use of numbered cores in accordance with BS 5467 provided that the numbers 1, 2 and 3 shall signify phase conductors, and the number 0 the neutral conductor.

(v) For mineral-insulated cables, the application at terminations of tapes, sleeves or discs of the appropriate colours prescribed in Table 52A.

NOTE – Identification sleeves for cables should comply with BS 3858 where appropriate.

Bare conductors shall be made identifiable where necessary by the application of tapes, sleeves or discs of the appropriate colours prescribed in Table 52A, or by painting with those colours.

Any scheme of colouring used to identify switchboard busbars or poles shall comply with the requirements of Table 52A so far as these are applicable.

TABLE 52 A

Colour identification of cores of non-flexible cables and bare conductors for fixed wiring

NOTE – For armoured p.v.c.-insulated cables and paper-insulated cables, see Regulation 524–3 (ii) and (iii).

Function	Colour identification
Protective (including earthing) conductor	green-and-yellow
Phase of a.c. single-phase circuit	red (or yellow or blue*)
Neutral of a.c. single- or three-phase circuit	black
Phase R of 3-phase a.c. circuit	red
Phase Y of 3-phase a.c. circuit	yellow
Phase B of 3-phase a.c. circuit	blue
Positive of d.c. 2-wire circuit	red
Negative of d.c. 2-wire circuit	black
Outer (positive or negative) of d.c. 2-wire circuit derived from 3-wire system	red
Positive of 3-wire d.c. circuit	red
Middle wire of 3-wire d.c. circuit	black
Negative of 3-wire d.c. circuit	blue

*As alternatives to the use of red, if desired, in large installations, on the supply side of the final distribution board.

Flexible cables and flexible cords

524–4 Every core of a flexible cable or flexible cord shall be identifiable throughout its length as appropriate to its function, as indicated in Table 52B.

524–5 Flexible cables or flexible cords having the following core colours shall not be used: green alone; yellow alone; or any bicolour other than the colour combination green-and-yellow.(See Regulation 524–1).

TABLE 52B

Colour identification of cores of flexible cables and flexible cords

Number of cores	Function of core	Colour(s) of core
1	Phase Neutral Protective	Brown[1] Blue Green-and-yellow
2	Phase Neutral	Brown Blue[2]
3	Phase Neutral Protective	Brown[3] Blue[2] Green-and-yellow
4 or 5	Phase Neutral Protective	Brown or black[4] Blue[2] Green-and-yellow

(1) Or any other colour not prohibited by Regulations 524–1 and 524–5, except blue.

(2) The blue core may be used for functions other than the neutral in circuits which do not incorporate a neutral conductor, in which case its function shall be appropriately identified during installation; provided that the blue core shall not in any event be used as a protective conductor. If the blue core is used for other functions, the coding L1, L2, L3, or other coding where appropriate should be used.

(3) In three-core flexible cables or flexible cords not incorporating a green-and-yellow core, a brown core and a black core may be used as phase conductors.

(4) Where an indication of phase rotation is desired, or it is desired to distinguish the functions of more than one phase core of the same colour, this shall be by the application of numbered or lettered (not coloured) sleeves to the cores, preferably using the coding L1, L2, L3 or other coding where appropriate.

525 PREVENTION OF MUTUAL DETRIMENTAL INFLUENCE

Between low voltage circuits and circuits of other categories

525–1 Low voltage circuits shall be segregated from extra-low voltage circuits as required by Chapter 41.

525–2 Where an installation comprises circuits for telecommunication, fire-alarm or emergency lighting systems, as well as circuits operating at low voltage and connected directly to a mains supply system, precautions shall be taken, in accordance with Regulations 525–3 to 525–9, to prevent electrical contact (and, for fire-alarm circuits and emergency lighting circuits, physical contact) between the cables of the various types of circuit.

NOTES: 1 – See the definition in Part 2 for 'circuit' which gives details of the three categories used in Regulations 525–3 to 525–9.

2 – Where it is proposed to install cables of Category 1 circuits in the same cable enclosure or duct as cables of a telecommunication system which may be connected to lines provided by a public telecommunications system authority, the approval of that authority is necessary.

3 – Cables used to connect the battery chargers of self-contained luminaires to the normal mains circuit are not considered to be emergency lighting circuits.

525–3 Cables of Category 1 circuits shall not be drawn into the same conduit, duct or ducting as cables of Category 2 circuits, unless the latter cables are insulated in accordance with the requirements of these Regulations for the highest voltage present in the Category 1 circuits.

525–4 Cables of Category 1 circuits shall not in any circumstances be drawn into the same conduit, duct or ducting as cables of Category 3 circuits.

NOTE – BS 5266 recommends that cables of emergency lighting systems are segregated from the cables of any other circuits.

525–5 Where a common channel or trunking is used to contain cables of Category 1 and Category 2 circuits, all cables connected to Category 1 circuits shall be effectually partitioned from the cables of the Category 2 circuits, or alternatively the latter cables shall be insulated in accordance with the requirements of these Regulations for the highest voltage present in the Category 1 circuits. (See also Regulation 525–7).

525–6 Where Category 3 circuits are installed in a channel or trunking containing circuits of any other category, these circuits shall be segregated from the latter by continuous partitions such that the specified integrity of the Category 3 circuits is not reduced. These partitions shall also be provided at any common outlets in a trunking system accommodating Category 3 circuits and circuits of other categories. Where mineral-insulated cables are used for the Category 3 circuits such partitions are not normally required unless specified in BS 5266 or BS 5839. Where partitions are not used, the mineral-insulated cables shall be rated for exposed-to-touch conditions as given in Tables 9J1 and 9J3 of Appendix 9.

525–7 In conduit, duct, ducting or trunking systems, where controls or outlets for Category 1 and Category 2 circuits are mounted in or on common boxes, switchplates or blocks, the cables and connections of the two categories of circuit shall be partitioned by means of rigidly fixed screens or barriers.

NOTE – BS 5266 recommends such partitions between emergency lighting circuits and all other circuits.

525–8 Where cores of Category 1 and Category 2 circuits are contained in a common multicore cable, flexible cable or flexible cord, the cores of the Category 2 circuits shall be insulated individually or collectively as a group, in accordance with the requirements of these Regulations, for the highest voltage present in the Category 1 circuits, or alternatively shall be separated from the cores of the Category 1 circuits by an earthed metal braid of equivalent current-carrying capacity to that of the cores of the Category 1 circuits. Where terminations of the two categories of circuit are mounted in or on common boxes, switchplates or blocks, they shall be partitioned in accordance with Regulation 525–7 or alternatively be mounted on separate and distinct terminal blocks adequately marked to indicate their function.

525–9 Cores of Category 1 and Category 3 circuits shall not in any circumstances be contained in a common multicore cable, flexible cable or flexible cord.

NOTES: 1 – BS 5266 recommends that cores of emergency lighting circuits are not contained within the same multicore cable as cores of other circuits.

2 – Recommendations for the segregation of telecommunication wiring from the wiring of circuits operating at low voltage are contained in CP 327 Part 3 and CP 1020.

Between electrical services and exposed metalwork of other services

525–10 Metal sheaths and armour of all cables operating at low voltage, and metal conduits, ducts, ducting and trunking and bare protective conductors associated with such cables, which might otherwise come into fortuitous contact with other fixed metalwork shall be either effectually segregated therefrom, or effectually bonded thereto.

NOTES: 1 – The bonding of the main earthing terminal to the metalwork of gas and water services, required by Regulation 413–2, has no effect upon the need for compliance with Regulation 525–10 as to segregation or additional bonding elsewhere in the installations.

2 – Bonding to the metalwork of other services may require the permission of the undertakings responsible. Bonding to Post Office earth wires should be avoided, unless authorised by the Post Office. For Protective Multiple Earthing systems, special requirements for bonding are applicable.

3 – Segregation from fixed metalwork as described in Regulation 525–10 is desirable also for non-metal-sheathed cables operating at low voltage.

525–11 Electrical services shall not be installed in the same conduit, ducting or trunking as pipes or tubes of non-electrical services, e.g. air, gas, oil, or water. This requirement does not apply where the various services are under common supervision and it is confirmed that no mutual detrimental influence can occur.

NOTE – For information concerning ducts containing electrical and other services, see CP 413.

525–12 No cables shall be run in a lift (or hoist) shaft unless they form part of the lift installation as defined in BS 5655.

526 ACCESSIBILITY

526–1 Joints in non-flexible cables and joints between non-flexible and flexible cables or cords shall be accessible for inspection. However, this requirement shall not apply to joints –

(i) in cables buried underground, or

(ii) enclosed in building materials having the ignitability characteristic 'P' as specified in BS 476 Part 5, or

(iii) made by welding, soldering, brazing or compression and contained within an enclosure (such as a box) of material having the ignitability characteristic 'P' as specified in BS 476 Part 5.

The joints mentioned in item (ii), if inaccessible, shall not be made by means of mechanical clamps.

526–2 Inspection-type conduit fittings shall be so installed that they can remain accessible for such purposes as the withdrawal of existing cables or the installing of additional cables.

527 JOINTS AND TERMINATIONS

527–1 Every connection at a cable termination or joint shall be mechanically and electrically sound, be protected against mechanical damage and any vibration liable to occur, shall not impose any appreciable mechanical strain on the fixings of the connection, and shall not cause any harmful mechanical damage to the cable conductor. Joints in non-flexible cables shall be made by soldering, brazing, welding, or mechanical clamps, or be of the compression type. All mechanical clamps and compression-type sockets shall securely retain all the wires of the conductor.

527–2 Terminations and joints shall be appropriate to the size and type of conductor with which they are to be used.

527–3 Terminations and joints shall be suitably insulated for the voltage of the circuits in which they are situated.

527–4 Where a termination or joint in an insulated conductor, other than a protective conductor, is not made in an accessory or luminaire complying with the appropriate British Standard, it shall be enclosed in material having the ignitability characteristic 'P' as specified in BS 476 Part 5. Such an enclosure may be formed by part of an accessory or luminaire and a part of the building structure.

527–5 Cores of sheathed cables from which the sheath has been removed and non-sheathed cables at the terminations of conduit, ducting or trunking shall be enclosed as specified in Regulation 527–4. Alternatively the enclosure may be a box complying with BS 5733 or BS 4662 or other appropriate British Standard.

527–6 Every compression joint shall be of a type which has been the subject of a test certificate as described in BS 4579.

NOTE – The appropriate tools specified by the manufacturers of the joint connectors should be used.

527–7 Terminations of mineral-insulated cables shall be provided with sleeves having a temperature rating similar to that of the seals.

527–8 Cable glands shall securely retain without damage the outer sheath or armour of the cables. Mechanical cable glands for rubber- and plastics-insulated cables shall comply with BS 4121 or BS 6121 where appropriate.

527–9 Appropriate cable couplers shall be used for connecting together lengths of flexible cable or flexible cord. (See also Regulations 553–10 to 553–12).

527–10 Ends of lengths of conduit shall be free from burrs, and where they terminate at boxes, trunking and accessories not fitted with spout entries, shall be treated so as to obviate damage to cables.

527–11 Substantial boxes of ample capacity shall be provided at every junction involving a cable connection in a conduit system.

527–12 Every outlet for cables from a duct system or ducting system, every joint in such a system, and every joint between such a system and another type of duct, ducting, or conduit, shall be formed so that the joints are mechanically sound and that the cables drawn in are not liable to suffer damage.

528 FIRE BARRIERS

528–1 Where cables, conduits, ducts, ducting or trunking pass through fire-resistant structual elements such as floors and walls designated as fire barriers, the opening made shall be sealed according to the appropriate degree of fire resistance. In addition, where cables, conduits or conductors are installed in channels, ducts, ducting, trunking or shafts which pass through such elements, suitable internal fire-resistant barriers shall be provided to prevent the spread of fire. (See also Regulation 523–6).

529 SUPPORTS, BENDS AND SPACE FACTORS

Supports

529–1 Every cable and conductor used as fixed wiring shall be supported in such a way that it is not exposed to undue mechanical strain and so that there is no appreciable mechanical strain on the terminations of the conductors, account being taken of mechanical strain imposed by the supported mass of the cable or conductor itself.

NOTE – See Appendix 11 for guidance on methods of support.

529–2 All conduit, ducting and trunking shall be properly supported and of a type suitable for any risk of mechanical damage to which they may be liable in normal conditions of service or adequately protected against such damage. The method of support for rigid p.v.c. conduits shall allow for the longitudinal expansion and contraction of the conduits which may occur with variation of temperature under normal operating conditions.

Bends

529–3 The internal radius of every bend in a non-flexible cable shall be such as not to cause damage to the cable and, for British Standard cables of the types mentioned in Table 52C below, not less than the appropriate value stated in the table.

TABLE 52C

Minimum internal radii of bends in cables for fixed wiring

Insulation	Finish	Overall diameter	Factor to be applied to overall diameter+ of cable to determine minimum internal radius of bend.
Rubber or p.v.c. (circular, or circular stranded copper or aluminium conductors)	Non-armoured	Not exceeding 10 mm	3(2)*
		Exceeding 10 mm but not exceeding 25 mm	4(3)*
		Exceeding 25 mm	6
	Armoured	Any	6
P.V.C. (solid aluminium or shaped copper conductors)	Armoured or non-armoured	Any	8
Impregnated paper	Lead sheath	Any	12
Mineral	Copper or aluminium sheath with or without p.v.c. covering	Any	6

+For flat cables the factor is to be applied to the major axis.

*The figure in brackets relates to single-core circular conductors of stranded construction installed in conduit, ducting or trunking.

529–4 The use of solid (non-inspection) conduit elbows or tees shall be restricted to –

– locations at the ends of conduits immediately behind a luminaire, outlet box or conduit fitting of the inspection type, or

– one solid elbow located at a position not more than 500 mm from a readily accessible outlet box in a conduit run not exceeding 10m between two outlet points provided that all other bends on the conduit run are not more than the equivalent of one right angle.

529–5 The radius of every conduit bend other than a bend complying with BS 4568 or BS 4607 shall be such as to allow compliance with Regulation 529–3 for bends in cables and, in any event, the inner radius of the bend shall be not less than 2.5 times the outside diameter of the conduit.

529–6 Every bend in a closed duct or ducting shall be of an inner radius allowing compliance with Regulation 529–3.

Space factors

529–7 The number of cables drawn into, or laid in, an enclosure of a wiring system shall be such that no damage is caused to the cables or to the enclosure during their installation.

NOTE – Appendix 12 gives guidance on the determination of the number of cables meeting the requirements of Regulation 529–7 and the values for cable capacities of conduits given in that Appendix have been based on the application of an easy pull of cables into the conduits.

CHAPTER 53

SWITCHGEAR
(for protection, isolation, and switching)

CONTENTS

CHAPTER 53

SWITCHGEAR

(for protection, isolation, and switching)

530 COMMON REQUIREMENTS

530–1 Where an item of switchgear is required by these Regulations to disconnect all live conductors of a circuit, it shall be of a type such that the neutral conductor cannot be disconnected before the phase conductors and is reconnected at the same time as or before the phase conductors.

530–2 Single-pole switchgear shall not be connected in the neutral conductor of TN or TT systems.

530–3 Every fuse and circuit breaker shall be selected for a voltage not less than the maximum voltage difference (r.m.s. value, for a.c.) which normally develops under fault conditions.

531 DEVICES FOR PROTECTION AGAINST ELECTRIC SHOCK

Overcurrent protective devices

531–1 For TN and TT systems, every overcurrent protective device which is to be used also for protection against electric shock (indirect contact) shall be selected so that its operating time is –

(i) appropriate to the value of fault current that would flow in the event of a fault of negligible impedance between a phase conductor and exposed conductive parts, so that the permissible final temperature of the related protective conductor is not exceeded (see also Regulation 543–1), and

(ii) appropriate to compliance with the requirements of Regulation 413–4.

NOTE – The operating time of an overcurrent protective device used in this way depends upon the value of the fault loop impedance which, in TT systems, may be liable to considerable change with time, for example with seasonal variations. In the absence of reliable information in this respect, protection against electric shock in TT systems is preferably to be provided by means other than the overcurrent protective devices. The use of residual current devices is a preferred alternative.

531–2 For IT systems, where exposed conductive parts are connected together and overcurrent protective devices are to be used to provide protection against electric shock in the event of a second fault, the requirements for the protective devices are similar to those for TN systems, as specified in Regulation 531–1.

Residual current devices

531–3 Residual current devices shall be capable of disconnecting all the phase conductors of the circuit.

531–4 The magnetic circuit of the transformers of residual current devices shall enclose all the live conductors of the circuit. The corresponding protective conductor shall be outside the magnetic circuit.

531–5 The residual operating current of the protective device shall comply with the requirements of Regulation 413–6 as appropriate to the type of system earthing (see Appendix 3). Where the characteristics of the current-using equipment to be supplied can be determined, it shall be verified that the vectorial sum of the leakage currents on the part of the installation situated on the load side of the device is appropriate to the characteristics of the device; and the vectorial sum of the leakage currents in normal service of the various items of equipment supplied by that part of the installation shall be less than one half the nominal residual operating current of the device. For TT and IT systems, allowance shall be made for any likely variation of earthing resistance with time, for example with seasonal variations.

NOTE – For large installations, sub-division of the earthing arrangements is likely to be necessary because of the magnitude of the total inherent leakage of the cables and other equipment, and similar sub-division may be necessary for smaller installations where devices having a residual operating current of 30mA or less are used.

531–6 Where the operation of a residual current device relies upon a separate auxiliary supply external to the device, then either –

(i) the device shall be of a type that will operate automatically in case of failure of the auxiliary supply, or

(ii) the device shall incorporate, or be provided with, a supply which shall be available automatically upon failure of the auxiliary supply.

531–7 Residual current devices shall be located outside the magnetic field of other equipment, unless it is verified that their operation will not be impaired thereby.

531–8 Where a residual current device for protection against indirect contact is used with, but separately from, overcurrent protective devices, it shall be verified that the residual current operated device is capable of withstanding, without damage, the thermal and mechanical stresses to which it is likely to be subjected in case of a short circuit occurring on the load side of the point at which it is installed.

NOTE – The stresses mentioned in Regulation 531–8 depend upon the prospective short circuit current at the point where the residual current device is installed, and the operating characteristics of the device providing short circuit protection.

Fault-voltage operated protective devices

531–9 The characteristics of every fault-voltage operated protective device shall be such as to comply with Regulation 413–3 for automatic disconnection in the event of a fault of negligible impedance between a phase conductor and exposed conductive parts, taking into account the impedances of the fault current loop at every point at which this method of protection is to be applied. For TT systems, allowances shall be made for any likely increase in the value of earthing resistance with time, for example, with seasonal variations (see also Section 544).

532 *(Reserved for future use)*

533 OVERCURRENT PROTECTIVE DEVICES

533–1 For every fuse and circuit breaker there shall be provided on or adjacent to it an indication of its intended nominal current as appropriate to the circuit it protects. For a semi-enclosed fuse the intended nominal current to be indicated is the value to be selected in accordance with Regulation 533–4.

NOTE – Fuses having links likely to be replaced by persons other than skilled persons or instructed persons should preferably be of a type such that a fuse link cannot inadvertently be replaced by a fuse link having a higher nominal current.

533–2 Fuses for overload protection having links likely to be replaced by persons other than skilled persons or instructed persons shall either –

(i) have marked on or adjacent to them an indication of the type of link intended to be used, or

(ii) be of a type such that there is no possibility of inadvertent replacement by a link having the intended nominal current but a higher fusing factor than that intended.

533–3 Fuses which are likely to be removed or replaced whilst the circuit they protect is energised, shall be of a type such that they can be thus removed or replaced without danger.

533–4 Fuses shall preferably be of the cartridge type. Where semi-enclosed fuses are selected, they shall be fitted with elements in accordance with the manufacturer's instructions if any; in the absence of such instructions, they shall be fitted with single elements of plain or tinned copper wire of the appropriate diameter specified in Table 53A.

TABLE 53A

Sizes of fuse elements of plain or tinned copper wire, for use in semi-enclosed fuses

Nominal current of fuse	Nominal diameter of wire
A	mm
3	0.15
5	0.2
10	0.35
15	0.5
20	0.6
25	0.75
30	0.85
45	1.25
60	1.53
80	1.8
100	2.0

533–5 Where circuit breakers may be operated by persons other than skilled or instructed persons, they shall be designed or installed so that it is not possible to modify the setting or the calibration of their overcurrent releases without a deliberate act involving either the use of a key or tool or a visible indication of their setting or calibration.

533–6 To avoid unnecessary interruption of supplies (see also Section 314), the characteristics and settings of devices for overcurrent protection shall be such that proper discrimination in their operation is ensured.

534 *(Reserved for future use)*

535 *(Reserved for future use)*

536 *(Reserved for future use)*

537 ISOLATING AND SWITCHING DEVICES

General

537–1 Isolating and switching devices shall comply with the appropriate requirements of the following Regulations 537–2 to 537–19. A common device may be used for more than one of these functions if the appropriate requirements for each function are met.

Devices for isolation

537–2 Devices for isolation shall effectively disconnect all live supply conductors from the circuit concerned, taking into account Regulation 460–2.

537–3 The isolating distances between contacts or other means of isolation when in the open position shall be not less than those specified for isolators (disconnectors) according to BS 5419.

537–4 Semiconductor devices shall not be used as isolators.

537–5 The position of the contacts or other means of isolation shall be either externally visible or clearly and reliably indicated. An indication of the isolated position shall occur only when the specified isolating distance has been attained in each pole.

537–6 Devices for isolation shall be selected and/or installed in such a way as to prevent unintentional reclosure.

NOTE – Such reclosure might be caused, for example, by mechanical shock and vibration.

Provision shall be made for securing off-load devices for isolation against inadvertent and unauthorised operation.

537–7 Isolation shall be achieved preferably by the use of a multipole device cutting off all poles of the supply or alternatively by the use of single-pole devices which are situated adjacent to each other. This requirement does not apply to neutral links in TN-S systems.

Devices for switching off for mechanical maintenance

537–8 Devices for switching off for mechanical maintenance shall be inserted where practicable in the main supply circuit. Alternatively, such devices may be inserted in the control circuit, provided that supplementary precautions are taken to provide a degree of safety equivalent to that of interruption of the main supply, e.g. where such an arrangement is specified in the appropriate British Standard.

537–9 Devices for switching off for mechanical maintenance or control switches for such devices shall be manually initiated and shall have an externally visible contact gap or a clearly and reliably indicated OFF or OPEN position. Indication of that position shall occur only when the OFF or OPEN position on each pole has been attained.

537–10 Devices for switching off for mechanical maintenance shall be selected and/or installed in such a way as to prevent unintentional reclosure.

NOTE – Such reclosure might be caused, for example, by mechanical shock and vibration.

537–11 Where switches are used as devices for switching off for mechanical maintenance, they shall be capable of cutting off the full load current of the relevant part of the installation.

Devices for emergency switching

537–12 Means of interrupting the supply for the purpose of emergency switching shall be capable of cutting off the full load current of the relevant part of the installation. Where appropriate, due account shall be taken of stalled motor conditions.

537–13 Means for emergency switching shall consist of –

– a single switching device directly cutting off the incoming supply, or

– a combination of several items of equipment operated by one initiation only and resulting in the removal of the hazard by cutting off the appropriate supply; emergency stopping may include the retention of supply for electric braking facilities.

Plugs and socket outlets shall not be used for emergency switching.

537–14 Devices for emergency switching shall be, where practicable, manually operated, directly interrupting the main circuit. Devices such as circuit breakers and contactors operated by remote control shall open on de-energisation of the coils, or other techniques of suitable reliability shall be employed.

537–15 The operating means (such as handles and pushbuttons) for devices for emergency switching shall be clearly identifiable and preferably coloured red. They shall be installed in a readily accessible position where the hazard might occur and, where appropriate, at any additional remote position from which the device for emergency switching may need to be operated.

537–16 The operating means of the device for emergency switching shall be of the latching type or capable of being restrained in the OFF or STOP position. A device in which the operating means automatically resets is permitted where both that operating means and the means of reenergising are under the control of one and the same person. The release of the emergency switching device shall not reenergise the equipment concerned, unless an appropriate warning that the equipment may restart automatically is clearly indicated.

537–17 Every fireman's emergency switch provided for compliance with Regulation 476–12 shall –

(i) be coloured red and have fixed on or near it a nameplate marked with the words 'FIREMAN'S SWITCH', the plate being the minimum size 150mm by 100mm, in lettering easily legible from a distance appropriate to the site conditions but not less than 13mm high, and

(ii) have its ON and OFF positions clearly indicated by lettering legible to a person standing on the ground at the intended site, and the OFF position shall be at the top, and

(iii) preferably, be provided with a lock or catch to prevent the switch being inadvertently returned to the ON position.

NOTE – It is desirable that the nameplate mentioned in Item (i) above be marked also with the name of the company which installed or (if different) which maintains the installation concerned.

Devices for functional switching

537–18 Plugs and socket outlets of rating not exceeding 16A may be used as switching devices but see also Regulation 537–13. A plug and socket outlet of rating exceeding 16A a.c. may be used as a switching device (other than an emergency switching device) where the plug and socket outlet has a breaking capacity appropriate to the use intended. Plugs and socket outlets of rating exceeding 16A shall not be used as switching devices for d.c. circuits.

537–19 Every switch for a discharge lighting circuit shall be designed and marked for such purposes in accordance with BS 3676. Alternatively, it shall have a nominal current not less than twice the total steady current which it is required to carry or, if used to control both filament lighting and discharge lighting, shall have a nominal current not less than the sum of the current of the filament lamps and twice the total steady current of the discharge lamps.

CHAPTER 54

EARTHING ARRANGEMENTS AND PROTECTIVE CONDUCTORS

CONTENTS

541	**GENERAL**
542	**CONNECTIONS TO EARTH**
542–1 to **542–9**	Earthing arrangements
542–10 to **542–15**	Earth electrodes
542–16 to **542–18**	Earthing conductors
542–19 and **542–20**	Main earthing terminals or bars
543	**PROTECTIVE CONDUCTORS**
543–1 to **543–3**	Cross-sectional areas
543–4 to **543–14**	Types of protective conductor
543–15 to **543–19**	Preservation of electrical continuity of protective conductors
544	**EARTHING, AND PROTECTIVE CONDUCTORS, FOR FAULT-VOLTAGE OPERATED PROTECTIVE DEVICES**
545	*(Number reserved for future use)*
546	**COMBINED PROTECTIVE AND NEUTRAL (PEN) CONDUCTORS**
547	**PROTECTIVE BONDING CONDUCTORS**
547–1	General
547–2 and **547–3**	Main equipotential bonding conductors
547–4 to **547–7**	Supplementary bonding conductors

CHAPTER 54

EARTHING ARRANGEMENTS AND PROTECTIVE CONDUCTORS

NOTE – See Appendix 13 for explanatory diagram concerning earthing arrangements and protective conductors.

541 GENERAL

541–1 Every means of earthing and every protective conductor shall be selected and erected so as to satisfy the requirements of these Regulations for the safety and proper functioning of the associated equipment of the installation.

541–2 The earthing system of the installation may be subdivided, in which case each part thus divided shall comply with the requirements of this chapter.

NOTE – Means of earthing for lightning protection systems should comply with CP 326.

542 CONNECTIONS TO EARTH

Earthing arrangements

542–1 The consumer's main earthing terminal shall be connected to Earth by one of the methods described in Regulations 542–2 to 542–5, as appropriate to the type of system of which the installation is to form part (see Section 312) and in compliance with Regulations 542–6 to 542–8.

542–2 For TN-S systems the consumer's main earthing terminal shall be connected to the earthed point of the source of energy.

NOTE – Part of the connection may be formed by the supply undertaking's lines and equipment.

542–3 For TN-C-S systems, where Protective Multiple Earthing is provided by the supply undertaking (in accordance with conditions laid down by the Secretary of State for Energy or the Secretary of State for Scotland, and with the concurrence of the Post Office), means shall be provided for the consumer's main earthing terminal to be connected by the supply undertaking to the neutral of the source of energy.

NOTE – See Appendix 3 for information concerning the provision of a consumer's main earthing terminal by a supply undertaking.

542–4 For TT and IT systems, the consumer's main earthing terminal shall be connected via an earthing conductor to an earth electrode complying with Regulations 542–10 to 542–15.

542–5 For TN-C systems where the consumer's installation comprises earthed concentric wiring or another system of PEN conductors, the external conductor or PEN conductor shall be connected to the consumer's main earthing terminal. Where the supply is obtained from a privately owned transformer or convertor, the main earthing terminal shall be connected to the neutral of the supply. Where the supply is provided by a supply undertaking under the conditions mentioned in Regulation 546–1, means shall be provided for the main earthing terminal to be connected by the supply undertaking to the neutral of the supply.

542–6 The earthing arrangements may be used jointly or separately for protective or functional purposes according to the requirements of the installation.

542–7 The earthing arrangements shall be such that –

(i) the value of resistance from the consumer's main earthing terminal to the earthed point of the supply for TN systems, or to Earth for TT and IT systems, is in accordance with the protective and functional requirements of the installation, and expected to be continuously effective, and

(ii) earth fault currents and earth leakage currents likely to occur are carried without danger, particularly from thermal, thermomechanical and electromechanical stresses, and

(iii) they are adequately robust or have additional mechanical protection appropriate to the assessed conditions of external influence.

542–8 Precautions shall be taken against the risk of damage to other metallic parts through electrolysis.

542–9 Where a number of installations have separate earthing arrangements, any protective conductor running between any two of the separate installations shall either be capable of carrying the maximum fault current likely to flow through it, or be earthed within one installation only and insulated from the earthing arrangements of any other installation. In the latter circumstances, if the protective conductor forms part of a cable, the protective conductor shall be earthed only in the installation containing the associated protective device.

Earth electrodes

542–10 The following types of earth electrode are recognised for the purposes of these Regulations:

- earth rods or pipes,
- earth tapes or wires,
- earth plates,
- earth electrodes embedded in foundations,
- metallic reinforcement of concrete,
- metallic pipe systems, where not precluded by Regulations 542–14 and 542–15,
- lead sheaths and other metallic coverings of cables, where not precluded by Regulation 542–15,
- other suitable underground structures.

NOTES: 1 – The efficacy of an earth electrode depends on local soil conditions, and one or more earth electrodes suitable for the soil conditions and value of earth electrode resistance required should be selected.

2 – The value of earth electrode resistance may be calculated or measured. A method of measurement is described in item 4 of Appendix 15.

542–11 The type and embedded depth of earth electrodes shall be such that soil drying and freezing will not increase the resistance of the earth electrode above the required value.

542–12 The materials used and the construction of the earth electrodes shall be such as to withstand damage due to corrosion.

542–13 The design of the earthing arrangements shall take account of possible increase in earth resistance of earth electrodes due to corrosion.

542–14 The metalwork of public gas and water services shall not be used as a sole protective earth electrode. This requirement does not preclude the bonding of such metalwork as required by Regulation 413–2.

542–15 Lead sheaths and other metallic coverings of cables not liable to deterioration through excessive corrosion may be used as earth electrodes provided the consent of the owner of the cables is obtained and suitable arrangements exist for the owner of the electrical installation to be warned of any proposed changes to the cable that might affect its suitability as an earth electrode.

Earthing conductors

542–16 Every earthing conductor shall comply with Section 543 and in addition, where buried in the soil, shall have a cross-sectional area not less than that stated in Table 54A.

TABLE 54A

Minimum cross-sectional areas of buried earthing conductors

	Mechanically protected	Not mechanically protected
Protected against corrosion	as required by Regulation 543–1	$16mm^2$ copper $16mm^2$ coated steel
Not protected against corrosion	$25mm^2$ copper $50mm^2$ steel	$25mm^2$ copper $50mm^2$ steel

NOTE – For tape or strip conductors, the thickness should be adequate to withstand mechanical damage and corrosion.

542–17 Aluminium and copperclad aluminium conductors shall not be used for final connections to earth electrodes.

542–18 The connection of an earthing conductor to an earth electrode or other means of earthing shall be soundly made and electrically and mechanically satisfactory, and labelled in accordance with Regulation 514–7.

Main earthing terminals or bars

542–19 In every installation a main earthing terminal or bar shall be provided to connect the following conductors to the earthing conductor:

– the circuit protective conductors, and
– the main bonding conductors, and
– functional earthing conductors (if required).

542–20 Provision shall be made in an accessible position for disconnecting the main earthing terminal from the means of earthing, to permit measurement of the resistance of the earthing arrangements. This joint shall be such that it can be disconnected only by means of a tool, be mechanically strong and reliably maintain electrical continuity.

543 PROTECTIVE CONDUCTORS

Cross-sectional areas

543–1 The cross-sectional area of every protective conductor, other than an equipotential bonding conductor, shall be –

(i) calculated in accordance with Regulation 543–2, or
(ii) selected in accordance with Regulation 543–3.

If the protective conductor is separate (i.e. does not form part of a cable, and is not formed by conduit, ducting or trunking, and is not contained in an enclosure formed by a wiring system), the cross-sectional area shall in any event be not less than –

– 2.5mm² if mechanical protection is provided.

– 4.0mm² if mechanical protection is not provided.

For an earthing conductor buried in the ground Regulation 542–16 also applies.

The cross-sectional areas of equipotential bonding conductors shall comply with Section 547.

543–2 The cross-sectional area, where calculated, shall be not less than the value determined by the following formula (applicable only for disconnection times not exceeding 5 seconds):

$$S = \frac{\sqrt{I^2 t}}{k} \text{ mm}^2$$

where
S is the cross-sectional area in mm²,

I is the value (r.m.s., for a.c.) of fault current for a fault of negligible impedance, which can flow through the associated protective device, in amperes,

NOTE – Account should be taken of the current limiting effect of the circuit impedances and the limiting capability (I^2t) of the protective device.

t is the operating time of the disconnecting device in seconds, corresponding to the fault current I amperes,
k is a factor dependent on the material of the protective conductor, the insulation and other parts, and the initial and final temperatures.

Values of k for protective conductors in various use or service are as given in Tables 54B, 54C, 54D and 54E. The values are based on the initial and final temperatures indicated below each table.

Where the application of the formula produces a non-standard size, a conductor of the nearest larger standard cross-sectional area shall be used.

NOTE – See Appendix 8 for further guidance on calculation of cross-sectional area of protective conductors.

TABLE 54B

Values of k for insulated protective conductors not incorporated in cables, or bare protective conductors in contact with cable covering.

Material of conductor	Insulation of protective conductor or cable covering		
	PVC	85°C rubber	90°C thermosetting
Copper	143	166	176
Aluminium	95	110	116
Steel	52	60	64
Assumed initial temperature	30°C	30°C	30°C
Final temperature	160°C	220°C	250°C

TABLE 54C

Values of k for protective conductor as a core in a cable

Material of conductor	Insulation material		
	PVC	85°C rubber	90°C thermosetting
Copper	115	134	143
Aluminium	76	89	94
Assumed initial temperature	70°C	85°C	90°C
Final temperature	160°C	220°C	250°C

TABLE 54D

Values of k for protective conductor as a sheath or armour of a cable

Material of conductor	Insulation material		
	PVC	85°C rubber	90°C thermosetting
Steel	44	51	54
Aluminium	81	93	98
Lead	22	26	27
Assumed initial temperature	60°C	75°C	80°C
Final temperature	160°C	220°C	250°C

TABLE 54E

Values of k for bare conductors where there is no risk of damage to any neighbouring material by the temperatures indicated.

Material of conductor	Conditions		
	Visible and in restricted areas*	Normal conditions	Fire risk
Copper	228	159	138
Aluminium	125	105	91
Steel	82	58	50
Assumed initial temperature	30°C	30°C	30°C
Final temperatures:			
Copper conductors	500°C	200°C	150°C
Aluminium conductors	300°C	200°C	150°C
Steel conductors	500°C	200°C	150°C

***The temperatures indicated are valid only where they do not impair the quality of the connections.**

543–3 Where it is desired not to calculate the minimum cross-sectional area of a protective conductor in accordance with Regulation 543–2, the value of cross-sectional area may be selected in accordance with Table 54F as follows:

where the protective conductor is made of the same material as the associated phase conductors, by selection of the appropriate value of Sp from the table, or

where the protective conductor and the associated phase conductors are made of different materials, by selection of the appropriate value of conductance not less than the conductance resulting from the application of the table.

Where the application of Table 54F produces a non-standard size, a conductor having the nearest standard cross-sectional area shall be used.

TABLE 54F

Minimum cross-sectional area of protective conductors in relation to the area of associated phase conductors

Cross-sectional area of phase conductor (S)	Minimum cross-sectional area of the corresponding protective conductor (Sp)
mm^2	mm^2
$S \leqslant 16$	S
$16 < S \leqslant 35$	16
$S > 35$	$\frac{S}{2}$

Types of protective conductor

543–4 Every protective conductor shall comply with the appropriate requirements of Chapter 52 and the following Regulations 543–5 to 543–14.

543–5 The protective conductor may form part of the same cable as the associated live conductors, in which case it shall comply with the applicable requirements of the appropriate British Standard for the cable (see Regulations 521–1 and 521–5).

543–6 The circuit protective conductor of every ring final circuit (other than that formed by the metal covering or enclosure of a cable) shall also be run in the form of a ring having both ends connected to the earth terminal at the origin of the circuit.

543–7 Enclosures of factory built assemblies, or metal-enclosed busbar trunking systems, where used as protective conductors, shall satisfy all the three following requirements:

(i) their electrical continuity shall be achieved in such a manner as to afford protection against mechanical, chemical or electrochemical deterioration,

(ii) their cross-sectional area shall be at least equal to that resulting from the application of Regulation 543–2, and

(iii) they shall permit the connection of other protective conductors at every predetermined tap-off point.

543–8 Every protective conductor formed wholly or partly by the metal sheath and/or armour or other metallic covering forming part of a cable shall comply with the requirements of Items (i) and (ii) of Regulation 543–7.

543–9 Metal enclosures for cables, where used as protective conductors, shall satisfy all the three following requirements:

(i) their cross-sectional area shall be at least equal to that resulting from the application of Regulation 543–2,

(ii) any joints along their run shall be made in accordance with Regulations 543–15 to 543–19, and

(iii) provision shall be made for the connection to them of other protective conductors as necessary.

543–10 Where the protective conductor is formed by conduit, trunking, ducting, or the metal sheath and/or armour of cables, the earthing terminal of each socket outlet shall be connected by a separate protective conductor to an earthing terminal incorporated in the associated box or other enclosure.

543–11 Flexible or pliable conduit shall not be used as a protective conductor.

543–12 Exposed conductive parts of equipment shall not be used to form part of protective conductors for other equipment except as provided by Regulations 543–7 to 543–10.

543–13 Metal enclosures for cables shall not be used as a PEN conductor.

543–14 Extraneous conductive parts shall not be used as protective conductors, except as permitted by Regulation 547–7.

Preservation of electrical continuity of protective conductors

543–15 Protective conductors shall be suitably protected against mechanical and chemical deterioration and electrodynamic effects. Every protective conductor not forming part of a cable or cable enclosure and having a cross-sectional area up to and including $6mm^2$, shall be protected throughout by insulation at least equivalent to that provided for a single-core non-sheathed cable of appropriate size complying with BS 6004; this requirement does not apply to copper strip. Where the sheath of a cable incorporating an uninsulated protective conductor is removed adjacent to joints and terminations, protective conductors of cross-sectional area up to and including $6mm^2$ shall be protected by insulating sleeving complying with BS 2848.

543–16 Connections of protective conductors shall comply with the requirements of Regulation 526–1 for accessibility, as though that regulation applied to protective conductors of all types. Those requirements however shall not apply to joints in metal conduit, ducting or trunking systems.

543–17 No switching device shall be inserted in a protective conductor, but joints which can be disconnected for test purposes by use of a tool are permitted.

543–18 Where electrical earth monitoring is used, the operating coil shall be connected in the pilot conductor and not in the protective earthing conductor.

543–19 All joints in metallic conduit shall be made mechanically and electrically continuous by screwing or by substantial mechanical clamps. Plain slip or pin-grip sockets shall not be used.

544 EARTHING, AND PROTECTIVE CONDUCTORS, FOR FAULT-VOLTAGE OPERATED PROTECTIVE DEVICES

544–1 Where protection for safety is afforded by a fault-voltage operated protective device, the requirements of Regulations 544–2 to 544–5 shall be observed.

544–2 An independent earth electrode shall be provided outside the resistance area of any other parallel earth. If by sub-division of the earthing system discrimination in operation between a number of fault-voltage operated protective devices is to be afforded, the resistance areas of the associated earth electrodes shall not overlap.

544–3 The voltage-sensitive element of the protective device shall be connected between the main earthing terminal and the earthing conductor.

544–4 The earthing conductor shall be insulated to avoid contact with other protective conductors or any exposed conductive parts or extraneous conductive parts, so as to prevent the voltage-sensitive element from being inadvertently bridged.

544–5 Protective conductors shall be connected only to the exposed conductive parts of those items of equipment whose supply is to be interrupted in the event of operation of the fault-voltage operated protective device, and to any extraneous conductive parts in the same equipotential zone.

545 *(Number reserved for future use)*

546 COMBINED PROTECTIVE AND NEUTRAL (PEN) CONDUCTORS

546–1 The provisions of this section may be applied only –

– where the supply undertaking concerned has been specially authorised in respect of the installation in question by the Secretary of State for Energy or the Secretary of State for Scotland (or, outside England, Wales and Scotland, by the corresponding authority) to permit additional connections of the neutral conductor to Earth, and where the installation complies with the conditions for that authorisation, or

– where the installation is supplied by a privately owned transformer or convertor in such a way that there is no metallic connection with the general public supply, or

– where the supply is obtained from a private generating plant.

546–2 Conductors of the following types may serve as PEN conductors provided that the part of the installation concerned is not supplied through a residual current device:

(i) for fixed installations, conductors of cables not subject to flexing and having a cross-sectional area not less than $10mm^2$,

(ii) the outer conductor of concentric cables where that conductor has a cross-sectional area not less than $4mm^2$ in a cable complying with an appropriate British Standard and selected and erected in accordance with Regulations 546–3 to 546–8.

546–3 The outer conductor of a concentric cable shall not be common to more than one circuit. This requirement does not preclude the use of a twin or multicore cable to serve a number of points contained within one final circuit.

546–4 The conductance of the outer conductor of a concentric cable (measured at a temperature of 20°C) shall –

(i) for a single-core cable, be not less than that of the internal conductor,

(ii) for a multicore cable in a multiphase or multipole circuit, be not less than that of one internal conductor,

(iii) for a twin or multicore cable serving a number of points contained within one final circuit or having the internal conductors connected in parallel, be not less than that of the internal conductors connected in parallel.

546–5 At every joint in the outer conductor of a concentric cable and at terminations, the continuity of that conductor shall be ensured by a conductor additional to any means used for sealing and clamping the outer conductor. The conductance of the additional conductor shall be not less than that specified in Regulation 546–4 for the outer conductor.

546–6 No means of isolation or switching shall be inserted in the outer conductor of a concentric cable.

546–7 The PEN conductor of every cable shall be insulated or have an insulating covering suitable for the highest voltage to which it may be subjected.

546–8 If from any point of the installation the neutral and protective functions are provided by separate conductors, those conductors shall not then be connected together beyond that point. At the point of separation, separate terminals or bars shall be provided for the protective and neutral conductors. The PEN conductor shall be connected to the terminal or bar intended for the protective earthing conductor.

547 PROTECTIVE BONDING CONDUCTORS

General

547–1 Aluminium or copperclad aluminium conductors shall not be used for bonding connections to water pipes likely to be frequently subjected to condensation in normal use.

Main equipotential bonding conductors

547–2 Main equipotential bonding conductors shall have cross-sectional areas not less than half the cross-sectional area of the earthing conductor of the installation, subject to a minimum of $6mm^2$. Except where

PME conditions apply, the cross-sectional area need not exceed 25mm^2 if the bonding conductor is of copper or a cross-sectional area affording equivalent conductance in other metals.

NOTE – Where PME conditions apply the supply undertaking should be consulted to determine any special requirements concerning the size of protective conductors.

547–3 Main equipotential bonding connections to any gas or water services shall be made as near as practicable to the point of entry of those services into the premises; provided that where there is an insulating section or insert at that point, the connection shall be made to the metalwork on the consumer's side of that section or insert.

For a gas service the bonding connection shall be made on the consumer's side of the meter, between the meter outlet union and any branch pipework.

NOTE – It is recommended that this connection should be made within 600mm of the gas meter.

Supplementary bonding conductors

547–4 A supplementary bonding conductor connecting two exposed conductive parts shall have a cross-sectional area not less than that of the smaller protective conductor connected to the exposed conductive parts, subject to a minimum of 2.5mm^2 if mechanical protection is provided, or 4mm^2 if mechanical protection is not provided.

547–5 A supplementary bonding conductor connecting exposed conductive parts to extraneous conductive parts shall have a cross-sectional area not less than that of the protective conductor connected to the exposed conductive part, subject to a minimum of 2.5mm^2 if mechanical protection is provided, or 4mm^2 if mechanical protection is not provided.

547–6 A supplementary bonding conductor connecting two extraneous conductive parts shall have a cross-sectional area not less than 2.5mm^2 if mechanical protection is provided or 4mm^2 if mechanical protection is not provided, except that where one of the extraneous conductive parts is connected to an exposed conductive part in compliance with Regulation 547–5, that Regulation shall apply also to the conductor connecting the two extraneous conductive parts.

547–7 Supplementary bonding may be provided by suitable extraneous conductive parts of a permanent and reliable nature, or by supplementary conductors, or a combination of these.

CHAPTER 55

OTHER EQUIPMENT

CONTENTS

551	**TRANSFORMERS**
552	**ROTATING MACHINES**
553	**ACCESSORIES**
553–1 to **553–9**	Plugs and socket outlets
553–10 to **553–12**	Cable couplers
553–13	Caravan couplers
553–14 to **553–18**	Lampholders
553–19 and **553–20**	Ceiling roses
554	**CURRENT-USING EQUIPMENT**
554–1 and **554–2**	Luminaires
554–3 to **554–19**	High voltage discharge lighting installations
554–20 to **554–26**	Electrode water heaters and boilers
554–27	Heaters for water or other liquids having immersed heating elements
554–28 to **554–30**	Water heaters having immersed and uninsulated heating elements
554–31 to **554–34**	Conductors and cables for soil, road and floor warming
554–35 to **554–40**	Electric fence controllers

CHAPTER 55

OTHER EQUIPMENT

551 TRANSFORMERS

551–1 Where an autotransformer is connected to a circuit having a neutral conductor, the common terminal of the winding shall be connected to the neutral conductor.

NOTE – Regulation 551–1 does not preclude the use of autotransformer starters complying with BS 4941 Part 4.

551–2 Step-up autotransformers shall not be connected to IT systems.

551–3 Where a step-up transformer is used, a linked switch shall be provided for disconnecting the transformer from all poles of the supply, including the neutral conductor.

551–4 Every inductor and high-reactance transformer for discharge lighting shall be placed as near as practicable to its associated discharge lamp.

551–5 Transformers for high voltage discharge lighting installations shall comply with the relevant regulations of Section 554.

552 ROTATING MACHINES

552–1 All equipment, including cables, of every circuit carrying the starting, accelerating and load currents of a motor shall be suitable for a current at least equal to the full-load current rating of the motor when rated in accordance with the appropriate British Standard. Where the motor is intended for intermittent duty and for frequent stopping and starting, account shall be taken of any cumulative effects of the starting periods upon the temperature rise of the equipment of the circuit.

NOTES: 1 – The supply undertaking should be consulted regarding starting arrangements for motors requiring heavy starting currents.

2 – For further guidance on the starting performance of motors, see BS 5000 Part 99.

552–2 The rating of circuits supplying rotors of slip-ring or commutator induction motors shall be suitable for the starting and load conditions.

552–3 Every electric motor having a rating exceeding 0.37 kW shall be provided with control equipment incorporating means of protection against overcurrent in the motor. This requirement does not apply to a motor incorporated in an item of current-using equipment complying as a whole with an appropriate British Standard.

552–4 Every motor shall be provided with means to prevent automatic restarting after a stoppage due to drop in voltage or failure of supply, where unexpected restarting of the motor might cause danger. This requirement does not apply where the failure of the motor to start after a brief interruption of the supply would be likely to cause greater danger. This requirement does not preclude arrangements for starting a motor at intervals by an automatic control device, where other adequate precautions are taken against danger from unexpected restarting.

553 ACCESSORIES

Plugs and socket outlets

553–1 Every plug and socket outlet shall comply with all the requirements of Items (i) to (iii) below, and in addition shall comply with the appropriate requirements of Regulations 553–2 to 553–8.

(i) It shall not be possible for any pin of the plug to be engaged with any live contact of its associated socket outlet while any other pin of the plug is completely exposed. This requirement does not apply to plugs and socket outlets for safety extra-low voltage circuits.

(ii) It shall not be possible for any pin of the plug to be engaged with any live contact of any socket outlet within the same installation other than the type of socket outlet for which the plug is designed.

(iii) For circuits of TN and TT systems, where the plugs are of a type containing a fuse, the plugs shall be non-reversible and so designed and arranged that no fuse can be connected in the neutral conductor.

553–2 Every plug and socket outlet shall be of the non-reversible type, with provision for the connection of a protective conductor. This requirement does not apply to plugs and socket outlets for safety extra-low voltage circuits, or for the special circuits mentioned in Regulation 553–5.

553–3 In low voltage circuits, plugs and socket outlets shall conform with the applicable British Standards listed in Table 55A, unless Regulation 553–5 is applicable.

TABLE 55A

Plugs and socket outlets for low voltage circuits

Type of plug and socket outlet	Rating (amperes)	Applicable British Standard
Fused plugs and shuttered socket outlets, 2-pole and earth, for a.c.	13	BS 1363 (fuses to BS 1362)
Plugs, fused or non-fused, and socket outlets, 2-pole and earth	2, 5, 15, 30	BS 546 (fuses, if any, to BS 646)
Plugs, fused or non-fused, and socket outlets, protected type, 2-pole with earthing contact	5, 15, 30	BS 196
Plugs and socket outlets (theatre type)	15	BS 1778
Plugs and socket outlets (industrial type)	16, 32, 63, 125	BS 4343

553–4 Socket outlets for household and similar use shall be of the shuttered type and, for a.c. installations, shall preferably be of a type complying with BS 1363.

553–5 Plugs and socket outlets other than those complying with BS 1363, 546, 196, 1778 or 4343, may be used in single-phase a.c. or two-wire d.c. circuits operating at a voltage not exceeding 250 volts for –

(i) the connection of electric clocks, provided that the plugs and socket outlets are designed specifically for that purpose, and that each plug incorporates a fuse of rating not exceeding 3 amperes complying with BS 646 or BS 1362 as appropriate.

(ii) the connection of electric shavers, provided that the socket outlets are either incorporated in a shaver unit complying with BS 3052 or, in rooms other than bathrooms, are a type complying with BS 4573.

(iii) circuits having special characteristics such that danger would otherwise arise.

553–6 Every plug and socket outlet for the connection of a caravan to a caravan site installation shall comply with BS 4343 and be of the two-pole and earthing contact type, with the earthing contact in the position specified in BS 4343 for an operating voltage of 220 to 240V (six o'clock position). The plug and socket outlet shall be of the splashproof type. The rated current of the plug and socket outlet shall be 16A, unless the maximum demand of the caravan necessitates the use of a higher rating. The socket outlets shall be so disposed that one is available within 20m of any position intended for the caravan intake.

553–7 Every plug and socket outlet for use in construction site installations shall comply with BS 4343. This requirement need not be observed for site offices, toilets and the like.

553–8 Where a socket outlet is mounted vertically on a wall or other structure, it shall be mounted at a height above the floor or any working surface such as to minimise the risk of mechanical damage to the socket outlet or to an associated plug and its flexible cord which might be caused during insertion, use or withdrawal of the plug.

553–9 Provision shall be made so that every portable appliance and portable luminaire can be fed from an adjacent and conveniently accessible socket outlet.

NOTE – Account should be taken of the length of flexible cord normally fitted to the majority of appliances and luminaires (i.e. 1.5m to 2m).

Cable couplers

553–10 Cable couplers shall comply where appropriate with BS 196, BS 1778, or BS 4343. Couplers shall be non-reversible with provision for the connection of a protective conductor. These requirements do not apply to cable couplers for safety extra-low voltage circuits.

553–11 Cable couplers shall be so arranged that the plug of the coupler is connected to the load side of the circuit.

553–12 Every cable coupler for use in construction site installations shall comply with BS 4343.

Caravan couplers

553–13 Every connector and inlet of a caravan for the intake of a low voltage supply shall comply with BS 4343 and be of the two-pole and earthing contact type, the other characteristics of the coupler being as specified in Regulation 553–6 for plugs and socket outlets. Every inlet shall be marked on the outside with an indication of the nominal voltage and shall be positioned in a suitable recess with a lid on the outside of the caravan.

Lampholders

553–14 Lampholders shall not be connected to any circuit where the rated current of the overcurrent protective device exceeds the appropriate value stated in Table 55B. This requirement does not apply where the lampholders and their wiring are enclosed in earthed metal or insulating material having the ignitability characteristic 'P' as specified in BS 476 Part 5 or where separate overcurrent protection is provided.

TABLE 55B

Overcurrent protection of lampholders

Type of lampholder (as designated in BS 5042)		Maximum rating (amperes) of over-current protective device protecting the circuit
Bayonet (BS 5042, Part 1):	B15	6
	B22	16
Edison screw (BS 5042, Part 2):	E14	6
	E27	16
	E40	16

553–15 Lampholders for filament lamps shall normally be used only in circuits operating at a voltage not exceeding 250 volts.

553–16 Bayonet lampholders B15 and B22 shall comply with BS 5042, Part 1 and shall have the temperature rating T2 described in that British Standard, unless the operating temperature of the associated lamp cap is expected to be less than 165°C in which case the lampholders may have the T1 temperature rating.

NOTE – For further guidance on the application of T1 and T2 temperature ratings, see Appendix A to BS 5042 Part 1.

553–17 Every lampholder in a room containing a fixed bath or shower cubicle shall comply with Regulation 471–38.

553–18 For circuits of TN or TT systems, the outer contact of Edison type screw or centre contact bayonet lampholders shall be connected to the neutral conductor.

Ceiling roses

553–19 A ceiling rose shall not be installed in any circuit operating at a voltage normally exceeding 250 volts.

553–20 A ceiling rose shall not be used for the attachment of more than one outgoing flexible cord unless it is specially designed for multiple pendants.

Luminaires

554–1 Where pendant luminaires are used, the associated accessories shall be suitable for the mass suspended (see also Regulation 523–32).

554–2 Luminaires in caravans shall preferably be fixed directly to the structure or lining of the caravan. Where a pendant luminaire is installed in a caravan, provision shall be made for securing the luminaire to prevent damage to the flexible cord or luminaire when the caravan is moved. Enclosed luminaires for filament lamps in caravans shall be so designed or mounted as to allow a free circulation of air between the fittings and the body of the caravan.

High voltage discharge lighting installations

554–3 No circuit for discharge lighting shall use a voltage exceeding 5kV r.m.s. to Earth, measured on open circuit.

554–4 Every discharge lighting circuit operating at a voltage exceeding low voltage and supplied from a transformer having a rated input exceeding 500 watts shall be provided with means for the automatic disconnection of the supply in the event of a short circuit or an earth leakage current which exceeds 20% of the normal steady current in the circuit.

554–5 Ancillary equipment for discharge lighting installations operating at voltages exceeding low voltage, including inductors, capacitors, resistors, and transformers, shall be either totally enclosed in a substantial earthed metal container (which may form part of a luminaire), or alternatively shall be placed in a suitably ventilated enclosure of material having the ignitability characteristic 'P' as specified in BS 476 Part 5 or of fire-resistant construction which is reserved for this equipment. A notice 'DANGER – HIGH VOLTAGE' shall be placed on every such container or enclosure. The word 'DANGER' shall be in block letters not less than 5mm high. The letters shall be in red on a white background and the size of each notice shall be not less than 65mm by 50mm overall.

554–6 No conductor which is in metallic connection with a discharge lamp operating at a voltage exceeding low voltage shall be in connection with any conductor of the mains supply other than by means of any connection with Earth complying with these Regulations; provided that autotransformers delivering a voltage not exceeding 1.5kV r.m.s. measured on open circuit may be used on a.c. 2-wire circuits in TN or TT systems, subject to the provision of means of isolation of both poles of the supply.

554–7 Means of isolation and switching for discharge lighting circuits shall comply with Sections 476 and 537.

554–8 Connections in discharge lighting circuits at a voltage exceeding low voltage shall be made by cables complying with BS 5055, except as provided in Regulations 554–9 and 554–10, and 554–18 and 554–19. Metal-sheathed, or armoured, or metal-sheathed and armoured, cable shall be used, except that cable having a non-metallic sheath may be used –

(i) in exterior installations, for inter-lamp series connections not exceeding 3m in length which are not likely to suffer mechanical damage, or which are installed in box signs constructed substantially of material having the ignitability characteristic 'P' as specified in BS 476 Part 5, and

(ii) in interior installations in self-contained luminaires.

For the purposes of this regulation, the term 'exterior installation' has the same significance as that defined in Regulation 476–12.

554–9 Subject to Regulation 554–10, bare or lightly insulated conductors of copper or nickel, having a cross-sectional area of not less than 0.4mm^2 (1/0.71 mm), may be used for series connections in circuits operating at a voltage exceeding low voltage, e.g. for window signs, provided that either –

(i) the conductor does not exceed 1m in length, is supported at intervals not greater than 500mm, is not exposed to the likelihood of mechanical damage, and is either completely protected by non-hygroscopic insulating material having the ignitability characteristic 'P' as specified in BS 476 Part 5 or by glass tubing. The glass tubing shall have a wall thickness not less than 1mm, an overall diameter not less than 5mm, and be so arranged as to be reasonably secure against any displacement which would expose any part of the live metal, or

(ii) the conductor is in an enclosure, the interior of which is accessible only to skilled persons.

554–10 For shopfront fascia installations, bare or lightly insulated conductors shall be used only for connections housed within an earthed metal enclosure or for connections between the terminals of electrode housings. For all other connections behind or through fascia panels, including series connections, armoured or metal-sheathed or metal-braided cable shall be used. In all instances the wiring shall comply with the requirements of BS 5055.

554–11 Cables and conductors in circuits operating at a voltage exceeding low voltage shall be supported at intervals not exceeding the appropriate value stated in Table 55C. Supports for insulated and braided cables and for bare conductors shall be of non-hygroscopic insulating material having the ignitability characteristic 'P' as specified in BS 476 Part 5.

TABLE 55C

Supports for cables and conductors in circuits operating at a voltage exceeding low voltage

	Spacing of supports (mm)	
Type of cable or conductor	Horizontal	Vertical
Bare conductors	500	500
Insulated-and-braided cables	500	800
Metal-sheathed, non-armoured cables	800	1250
Armoured, or metal-sheathed and armoured cables	1000	1500

554–12 The length in millimetres measured along its centre line, of every support which serves to separate bare metal or cables operating at a voltage exceeding low voltage, the cables being neither metal-sheathed nor armoured, from earthed metalwork, woodwork, or surfaces likely to become damp, shall be not less than the value obtained by multiplying by 10 the voltage to Earth of the transformer secondary in kilovolts (r.m.s.), measured on open circuit.

554–13 In circuits operating at a voltage exceeding low voltage the air gap in millimetres from metal or cables which are neither metal-sheathed nor armoured, to earthed metal, woodwork, or surfaces likely to become damp, shall be not less than the value obtained by multiplying by 4 the voltage to Earth of the transformer secondary in kilovolts (r.m.s.), measured on open circuit.

554–14 In a circuit operating at a voltage exceeding low voltage, every cable shall be supported close to each terminal connection and in no instance at a distance from the terminal greater than 150mm, or 300mm if a glazed porcelain electrode-receptacle forming an adequate support for the cable is used.

554–15 Where a connection is made to a cable in a circuit operating at a voltage exceeding low voltage, the insulation exposed by removing the metal sheath or braid shall be suitably protected from the effects of ozone.

554–16 Where likely to suffer mechanical damage, every cable in a circuit operating at a voltage exceeding low voltage shall be armoured or otherwise suitably protected. Non-armoured cables shall not be drawn into metal tubing, except where they pass through walls or floors where they may be installed in short lengths of metal conduit which shall be earthed.

554–17 Where not otherwise readily identifiable, cables in a circuit operating at a voltage exceeding low voltage or their protective covering shall be distinguished by tabs or labels marked 'DANGER' securely attached at intervals not greater than 1.5m. The letters shall be in red on a white background, and shall be not less than 10mm high.

554–18 The return cable from an electrode to a transformer terminal which is earthed may be a low voltage cable in accordance with the relevant British Standard provided that the cross-sectional area of the conductor is not less than: 2.5mm^2 if mechanical protection is provided or 4mm^2 if mechanical protection is not provided.

554–19 The metalwork of a rotating device may be used as a return conductor provided that adjacent uninsulated metalwork is permanently and effectively earthed.

Electrode water heaters and boilers

554–20 Electrode boilers and electrode water heaters shall be connected to a.c. systems only, and shall be selected and erected in accordance with the appropriate requirements of this section.

554–21 The supply to the heater or boiler shall be controlled by a circuit breaker which shall be of the multipole linked type arranged to disconnect the supply from all electrodes simultaneously and provided with overcurrent protective devices in each conductor feeding an electrode.

554–22 The earthing of the heater or boiler shall comply with the requirements of Chapter 54 and, in addition, the shell of the heater or boiler shall be bonded to the metallic sheath and armour, if any, of the incoming supply cable. The earthing conductor shall be connected to the shell of the heater or boiler and shall have a conductance not less than that of the largest phase conductor connected to the equipment or, where an earth-leakage protective device is provided, not less than that corresponding to the operating current of that device, subject to a minimum conductor size of $2.5mm^2$ if mechanical protection is provided, or $4mm^2$ if mechanical protection is not provided.

554–23 Where an electrode water heater or electrode boiler is directly connected to a supply at a voltage exceeding low voltage, the installation shall include a residual current device arranged to disconnect the supply from the electrodes on the occurrence of a sustained earth-leakage current in excess of 10% of the rated current of the heater or boiler under normal conditions of operation, except that if in any instance a higher value is essential to ensure stability of operation of the heater or boiler, the value may be increased to a maximum of 15%. A time delay may be incorporated in the device to prevent unnecessary operation in the event of imbalance of short duration.

554–24 Where an electrode water heater or electrode boiler is connected to a three-phase low-voltage supply, the shell of the heater or boiler shall be connected to the neutral of the supply as well as to the earthing conductor. The current-carrying capacity of the neutral conductor shall be not less than that of the largest phase conductor connected to the equipment.

NOTE – Attention is drawn to the requirements of Regulation 4 (vii) and (viii) of the Electricity Supply Regulations, 1937.

554–25 Where the supply to an electrode water heater or electrode boiler is single-phase and one electrode is connected to a neutral conductor earthed by the supply undertaking, the shell of the water-heater or boiler shall be connected to the neutral of the supply as well as to the earthing conductor. This regulation does not apply where the arrangement described in Regulation 554–26 is adopted.

554–26 Where the heater or boiler is not intended to be piped to a water supply or to be in physical contact with any earthed metal, and where the electrodes and the water in contact with the electrodes are so shielded in insulating material that it is impossible to touch either the electrodes or the water in contact therewith while the electrodes are live, a fuse in the phase conductor may be substituted for the circuit breaker required under Regulation 554–21 and the shell of the heater or boiler need not be connected to the neutral of the supply.

Heaters for water or other liquids having immersed heating elements

NOTE – In the selection of heaters or boilers due account should be taken of any aggressive conditions of water (or other liquid) to which they may be subjected.

554–27 Every heater for water or other liquid shall incorporate or be provided with an automatic device to prevent a dangerous rise in temperature.

Water heaters having immersed and uninsulated heating elements

554–28 Every single-phase water heater or boiler having an uninsulated heating element immersed in the water shall comply with the requirements of Regulations 554–29 and 554–30.

NOTE – This type of water heater or boiler is deemed not to be an electrode water heater or boiler (see definition).

554–29 All metal parts of the heater or boiler which are in contact with the water (other than current-carrying parts) shall be solidly and metallically connected to a metal water-pipe through which the water supply to the heater or boiler is provided, and that water-pipe shall be in effective electrical connection with Earth by a means independent of the circuit protective conductor.

554–30 The heater or boiler shall be permanently connected to the electricity supply through a double-pole linked switch which is either separate from and within easy reach of the heater or boiler or is incorporated therein subject to compliance with Regulation 476–8, and the wiring from the heater or boiler shall be directly connected to that switch without the use of a plug and socket outlet; and, where the heater or boiler is installed in a room containing a fixed bath, the switch shall comply in addition with Regulation 471–39.

NOTES:

1 – Before a heater or boiler of the type referred to in Regulations 554–28 to 554–30 is installed, it is essential that the installer should confirm that no single-pole switch, non-linked circuit breaker or fuse is fitted in the neutral conductor in any part of the circuit between the heater or boiler and the origin of the installation.

2 – The agreement of the water supply authority will be required, and the fact that the heating element is in contact with the water should be brought to their notice, before a heater or boiler of this type is installed.

3 – The use of a heater or boiler of this type is inadvisable where the water supply is obtained through a water-softener of the salt regenerative type.

Conductors and cables for soil, road and floor warming

554–31 Where heating cables are required to pass through, or be in close proximity to, materials which present a fire hazard, the cables shall be enclosed in material having the ignitability characteristic 'P' as specified in BS 476 Part 5 and shall be adequately protected from any mechanical damage that can be expected to occur during service.

554–32 Heating cables intended for laying directly in soil, concrete, cement screeds or other materials used for the construction of roads and buildings shall be –

(i) capable of withstanding mechanical damage under the conditions that can reasonably be expected to prevail during their installation, and

(ii) constructed of materials that will be resistant to damage from dampness and/or corrosion under normal conditions of service.

554–33 Heating cables laid directly in soil, roads, or the structure of buildings shall be installed so that they –

(i) are completely embedded in the substance they are intended to heat, and

(ii) do not suffer damage in the event of movements normally to be expected in them or the substance in which they are embedded.

554–34 The loading of every floor-warming cable under operating conditions shall be limited to a value such that the appropriate conductor temperature specified in Table 55D is not exceeded.

TABLE 55D

Maximum operating conductor temperatures for floor-warming cables

Type of cable	Maximum operating conductor temperature °C
General-purpose p.v.c. over conductor	70
Enamelled conductor, polychloroprene over enamel, p.v.c. overall	70
Enamelled conductor p.v.c. overall	70
Enamelled conductor, p.v.c. over enamel, lead-alloy 'E' sheath overall	70
Heat-resisting p.v.c. over conductor	85
Nylon over conductor, heat-resisting p.v.c. overall	85
Butyl rubber or equivalent elastomeric insulation over conductor	85
Mineral insulation over conductor, copper sheath overall	105 or 135*
Silicone-treated woven-glass sleeve over conductor	180

*The choice of temperature depends upon the type of seal employed. Where the cable is provided with an overall covering of p.v.c., the temperatures stated for p.v.c. in the Table are applicable.

Electric fence controllers

554–35 Mains-operated electric fence controllers shall comply with BS 2632.

554–36 Every mains-operated electric fence controller shall be so installed that, so far as is reasonably practicable, it is free from risk of mechanical damage or unauthorized interference.

554–37 A mains-operated electric fence controller shall not be fixed to any pole of an overhead power or telecommunication line; provided that, where a low voltage supply to an electric fence controller is carried by insulated overhead line from a distribution board, the controller may be fixed to the pole carrying the supply.

554–38 Any earth electrode connected to the earth terminal of an electric fence controller shall be separate from the earthing system of any other circuit and shall be situated outside the resistance area of any electrode used for protective earthing.

NOTE – A method of verifying that an earth electrode is outside the resistance area of another, is recommended in Item 4 of Appendix 15.

554–39 Not more than one controller shall be connected to any electric fence or similar system of conductors.

554–40 Every electric fence or similar system of conductors and the associated controller shall be so installed that it is not liable to come into contact with any power or telecommunication apparatus or wiring, including an overhead power line, telephone or telegraph wires or a radio aerial, or with a protective conductor.

NOTE – Where battery-operated electric fence controllers are used, it is essential that the battery be disconnected from the controller for charging.

PART 6
Inspection and testing

CONTENTS

PART 6

Inspection and testing

CHAPTER 61

INITIAL INSPECTION AND TESTING

611 GENERAL

611–1 Every installation shall, on completion and before being energised, be inspected and tested in accordance with the requirements of this chapter to verify, as far as practicable, that the requirements of these Regulations have been met.

The methods of test shall be such that no danger to persons or property or damage to equipment can occur even if the circuit tested is defective.

611–2 The information required by Regulation 514–3 shall be made available to the person or persons carrying out the inspection and testing.

612 VISUAL INSPECTION

612–1 A visual inspection shall be made to verify that the installed electrical equipment is –

– in compliance with the applicable British Standards (this may be ascertained by mark or by certification furnished by the installer or by the manufacturer), and

– correctly selected and erected in accordance with these Regulations, and

– not visibly damaged so as to impair safety.

NOTE – Appendix 14 lists the principal features of an installation to be visually inspected in order to verify compliance with the Regulations.

613 TESTING

General

613–1 The following items, where relevant, shall be tested in the sequence indicated. Standard methods of testing, in respect of some of the following regulations of this section, are given in Appendix 15; the use of other methods is not precluded provided that they give no less effective results.

– Continuity of ring final circuit conductors,
– Continuity of protective conductors, including main and supplementary equipotential bonding,
– Earth electrode resistance,
– Insulation resistance,
– Insulation of site-built assemblies,
– Protection by electrical separation,
– Protection by barriers or enclosures provided during erection,
– Insulation of non-conducting floors and walls,
– Polarity,
– Earth fault loop impedance,
– Operation of residual current devices and fault-voltage operated protective devices.

In the event of any test indicating failure to comply, that test and those preceding, the results of which may have been influenced by the fault indicated, shall be repeated after the fault has been rectified.

Continuity of ring final circuit conductors

613–2 A test shall be made to verify the continuity of all conductors (including the protective conductor) of every ring final circuit (see Item 2 of Appendix 15).

Continuity of protective conductors

613–3 Every protective conductor shall be separately tested to verify that it is electrically sound and correctly connected. This test shall include all conductors and any extraneous conductive parts used for equipotential bonding (see Item 3 of Appendix 15).

Earth electrode resistance

613–4 Where it is necessary to measure the resistance of an earth electrode, the test method described in Item 4 of Appendix 15 shall be used.

Insulation resistance

613–5 The tests described in Regulations 613–6 to 613–8 shall be made before the installation is permanently connected to the supply. For these tests large installations may be divided into groups of outlets, each containing not less than 50 outlets. For the purposes of this regulation the term 'outlet' includes every point and every switch except that a socket outlet, appliance or luminaire incorporating a switch is regarded as one outlet. A d.c. voltage not less than twice the nominal voltage of the circuit concerned (r.m.s. value for an a.c. supply) shall be applied for the measurement of insulation resistance, provided that the test voltage need not exceed 500V d.c. for installations rated up to 500V, or 1000V d.c. for installations rated above 500V up to 1000V.

613–6 When measured with all fuse links in place, all switches (including, if practicable, the main switch) closed and, except for TN-C systems, all poles or phases of the wiring electrically connected together, the insulation resistance to Earth shall be not less than 1 megohm.

613–7 When measured between all the conductors connected to any one phase or pole of the supply and, in turn, all conductors connected to each other phase or pole, the insulation resistance shall be not less than 1 megohm. Wherever practicable, so that all parts of the wiring may be tested, all lamps shall be removed and all current-using equipment shall be disconnected and all local switches controlling lamps or other equipment shall be closed. Where the removal of lamps and/or the disconnection of current-using equipment is impracticable, the local switches controlling such lamps and/or equipment shall be open. Particular attention shall be given to the presence of electronic devices connected in the installation and such devices shall be isolated so that they are not damaged by the test voltage.

613–8 Where equipment is disconnected for the tests prescribed in Regulations 613–6 and 613–7, and the equipment has exposed conductive parts required by these Regulations to be connected to protective conductors, the insulation resistance between the exposed conductive parts and all live parts of the equipment shall be measured separately and shall comply with the requirements of the appropriate British Standard for the equipment. If there is no appropriate British Standard the insulation resistance shall be not less than 0.5 megohm.

Insulation of site-built assemblies

613–9 Where protection against direct contact is intended to be afforded by insulation applied to live parts during erection in accordance with Regulation 412–2, it shall be verified that the insulation is capable of withstanding, without breakdown or flashover, an applied voltage test equivalent to that specified in the British Standard for similar factory-built equipment.

613–10 Where protection against indirect contact is provided by supplementary insulation applied to equipment during erection in accordance with Regulations 413–18 to 413–26, it shall be verified by test –

- that the insulating enclosure affords a degree of protection not less than IP2X (see BS 5490), and
- that the insulating enclosure is capable of withstanding, without breakdown or flashover, an applied voltage test equivalent to that specified in the British Standard for similar factory-built equipment.

Electrical separation of circuits

613–11 Where Regulation 411–6 or 413–35 applies, the electrical separation of the separated circuit shall be inspected and/or tested.

Protection against direct contact, by barriers or enclosures provided during erection

613–12 Where protection against direct contact is intended to be afforded by barriers or enclosures provided during erection in accordance with Regulations 412–3 to 412–6, it shall be verified by test that the enclosures or barriers afford a degree of protection not less than IP2X or IP4X as appropriate, where those regulations so require.

Insulation of non-conducting floors and walls

613–13 Where protection against indirect contact is to be provided by a non-conducting location intended to comply with Regulations 413–27 to 413–31 and 471–19, the resistance of the floors and walls of the location to the main protective conductor of the installation shall be measured at not less than three points on each relevant surface, one of which shall be not less than 1m and not more than 1.2m from any extraneous conductive part in the location.

Polarity

613–14 A test of polarity shall be made and it shall be verified that all fuses and single-pole control devices are connected in the phase conductor only, that centre-contact bayonet and Edison-type screw lampholders in circuits having an earthed neutral conductor have their outer or screwed contacts connected to that conductor, and that wiring has been correctly connected to socket outlets.

Earth fault loop impedance

613–15 Where protective measures are used which require a knowledge of earth fault loop impedance, the relevant impedances shall be measured, or determined by an equally effective method (see Item 5 of Appendix 15).

Operation of residual current operated and fault-voltage operated protective devices

613–16 Where protection against indirect contact is to be provided by a residual current device or a fault-voltage operated protective device, its effectiveness shall be verified by a test simulating an appropriate fault condition and independent of any test facility incorporated in the device (see Item 6 of Appendix 15).

614 CERTIFICATION

614–1 Following the inspection and testing required by this chapter, a completion certificate shall be given by the contractor or other person responsible, or by an authorised person acting on his behalf, to the person ordering the work. The certificate shall be in the form set out in Appendix 16. Any defects or omissions revealed by inspection or test shall be made good before a completion certificate is issued, unless Regulation 622–1 is applicable.

NOTES: 1 – An inspection certificate (see Regulation 631–1) should accompany and be attached to the completion certificate.

2 – It is recommended that the contractor or other person responsible for the construction of the installation should remind the consumer of the importance of periodic re-inspection and re-testing at the appropriate time. Every re-inspection of the installation should be reported upon by means of an inspection certificate which should be submitted to the consumer and be signed by a competent person who should preferably be one of the following:

– a professionally qualified electrical engineer,

– a member of the Electrical Contractors' Association,

– a member of the Electrical Contractors' Association of Scotland,

– an approved contractor of the National Inspection Council for Electrical Installation Contracting, or

– a qualified person acting on behalf of one of the above (in which event it should be stated for whom he is acting).

CHAPTER 62

ALTERATIONS TO INSTALLATIONS

621 GENERAL

621–1 For an alteration to an existing installation, it shall be verified that the alteration complies with these Regulations and does not impair the safety of the existing installation.

622 CERTIFICATION

622–1 The requirements of Regulation 614–1 for the issue of a completion certificate shall apply to all work of the alteration. Any defects or omissions revealed in that work shall be made good before a completion certificate is issued. The contractor or other person responsible for the new work, or an authorised person acting on his behalf, shall report to the person ordering the work any defects found in related parts of the existing installation.

CHAPTER 63

PERIODIC INSPECTION AND TESTING

631–1 The results of a periodic inspection and test of an installation, or any part thereof, shall be recorded on an inspection certificate and given by the contractor or by an authorised person acting on his behalf, to the person ordering the inspection and test. The methods of inspection and test shall be in accordance with the relevant requirements of Chapter 61, and the certificate shall be in the form set out in Appendix 16.

Appendices

CONTENTS

APPENDIX 1

PUBLICATIONS OF THE BRITISH STANDARDS INSTITUTION TO WHICH REFERENCE IS MADE IN THESE REGULATIONS

The following list is limited to those British Standard Codes of Practice and British Standard Specifications to which specific reference is made in these Regulations. Many other BSI publications are of interest to those concerned with electrical installations and the BSI Yearbook or Sectional Lists should be consulted.

Copies of BSI publications are obtainable from the British Standards Institution, Sales Department, Newton House, 101 Pentonville Road, London, N1 9ND.

Specifications and Codes of Practice prefixed by letters 'BS'

31	Steel conduit and fittings for electrical wiring.
67	Ceiling roses.
88	Cartridge fuses for voltages up to and including 1000V a.c. and 1500V d.c.
125	Hard-drawn copper and copper-cadmium conductors for overhead power transmission purposes.
196	Protected-type non-reversible plugs, socket-outlets, cable-couplers and appliance-couplers, with earthing contacts for single phase a.c. circuits up to 250 volts.
215	Aluminium conductors and aluminium conductors, steel-reinforced, for overhead power transmission.
476	Fire tests on building materials and structures.
546	Two-pole and earthing-pin plugs, socket-outlets and socket-outlet adaptors for circuits up to 250 volts.
646	Cartridge fuse-links (rated at up to 5 amperes) for a.c. and d.c. service.
731 Part 1	Flexible steel conduit and adaptors for the protection of electric cable.
1004	Zinc alloys for die casting and zinc alloy die castings.
1361	Cartridge fuses for a.c. circuits in domestic and similar premises.
1362	General purpose fuse links for domestic and similar purposes (primarily for use in plugs).
1363	13A plugs, switched and unswitched socket-outlets and boxes.
1710	Identification of pipelines.
1778	15A three-pin plugs, socket-outlets and connectors (theatre type).
2484	Cable covers, concrete and earthenware.
2632	Mains-operated electric fence controllers.
2754	Construction of electrical equipment for protection against electric shock.
2848	Flexible insulating sleeving for electrical purposes.
3036	Semi-enclosed electric fuses (ratings up to 100 amperes and 240 volts to earth).
3052	Electric shaver supply units.
3242	Aluminium alloy stranded conductors for overhead power transmission.
3456	Safety of household and similar electrical appliances.
3535	Safety isolating transformers for industrial and domestic purposes.

3676	Switches for domestic and similar purposes (for fixed or portable mounting).
3858	Binding and identification sleeves for use on electric cables and wires.
3871	Miniature and moulded case circuit-breakers.
3939	Graphical symbols for electrical power, telecommunications and electronics diagrams.
4099	Colours of indicator lights, push buttons, annunciators and digital readouts.
4121	Mechanical cable glands for rubber and plastics insulated cables.
4343	Industrial plugs, socket outlets and couplers for a.c. and d.c. supplies.
4533	Electric luminaires (lighting fittings).
4553	PVC-insulated split concentric cables with copper conductors for electric supply.
4568	Steel conduit and fittings with metric threads of ISO form for electrical installations.
4573	Two-pin reversible plugs and shaver socket-outlets.
4579	Performance of mechanical and compression joints in electric cable and wire connectors.
4607	Non-metallic conduits and fittings for electrical installations.
4662	Boxes for the enclosure of electrical accessories.
4678	Cable trunking.
4727	Glossary of electrotechnical, power, telecommunication, electronics, lighting and colour terms.
4752	Switchgear and controlgear for voltages up to and including 1000V a.c. and 1200V d.c.
4941	Motor starters for voltages up to and including 1000V a.c. and 1200V d.c.
5000	Rotating electrical machines of particular types or for particular applications.
5042	Lampholders and starterholders.
5055	PVC-insulated and elastomer-insulated cables for electric signs and high voltage luminous discharge-tube installations.
5266	Emergency lighting of premises.
5345	Selection, installation and maintenance of electrical apparatus for use in potentially explosive atmospheres, (other than mining applications or explosive processing and manufacture).
5419	Air-break switches, air-break disconnectors, air-break switch disconnectors and fuse combination units for voltages up to and including 1000V a.c. and 1200V d.c.
5467	Armoured cables with thermosetting insulation for electricity supply.
5486	Factory-built assemblies of switchgear and controlgear for voltages up to and including 1000V a.c. and 1200V d.c.
5490	Degrees of protection provided by enclosures.
5593	Impregnated paper-insulated cables with aluminium sheath/neutral conductor and three shaped solid aluminium phase conductors (CONSAC), 500/1000V, for electricity supply.
5655	Lifts and service lifts.
5733	General requirements for electrical accessories.
5839	Fire detection and alarm systems in buildings.

6004	PVC-insulated cables (non-armoured) for electric power and lighting.
6007	Rubber-insulated cables for electric power and lighting.
6081	Terminations for mineral insulated cables.
6116	Elastomer-insulated flexible trailing cables for quarries and miscellaneous mines.
6121	Mechanical cable glands for elastomer and plastics insulated cables.
6207	Mineral-insulated cables.
6231	PVC-insulated cables for switchgear and controlgear wiring.
6346	PVC-insulated cables for electricity supply.
6480	Impregnated paper-insulated cables for electricity supply.
6485	PVC-covered conductors for overhead power lines.
6500	Insulated flexible cords.
6977	Braided travelling cables for electric and hydraulic lifts.

Codes of Practice (prefixed by letters 'CP')

326	The protection of structures against lightning.
327	Telecommunication facilities in buildings.
413	Ducts for building services.
1003	Electrical apparatus and associated equipment for use in explosive atmospheres of gas or vapour other than mining applications.
1013	Earthing.
1017	Distribution of electricity on construction and building sites.
1018	Electric floor-warming systems for use with off-peak and similar supplies of electricity.
1020	The reception of sound and television broadcasting.

NOTE – The use of the letters 'CP' to denote a Code of Practice has been discontinued and such Codes published since 1975 have the prefix letters 'BS'.

APPENDIX 2

STATUTORY REGULATIONS AND ASSOCIATED MEMORANDA

1. In Great Britain the following classes of electrical installations are required to comply with the statutory regulations indicated below. The Regulations listed represent the principal legal requirements. Information concerning those regulations may be obtained from the appropriate authority also indicated below.

Provisions relating to electrical installations are also to be found in other legislation relating to particular activities.

(i)	Installations generally, subject to certain exemptions.	Electricity Supply Regulations, 1937.	Secretary of State for Energy and Secretary of State for Scotland.
(ii)	Buildings generally (for Scotland only), subject to certain exemptions.	Building Standards (Scotland) Regulations with Amendment Regulations 1971-1980.	Secretary of State for Scotland.
(iii)	Factory installations, installations on construction sites, installations of non-domestic caravans such as mobile workshops.	Electricity (Factories Act) Special Regulations 1908 and 1944.	Health and Safety Commission.
(iv)	Cinematograph installations.	Cinematograph Regulations made under the Cinematograph Act, 1909, and/or Cinematograph Act, 1952.	Home Office and Secretary of State for Scotland.
(v)	Coal mine installations, (including stratified ironstone, shale or fireclay).	Coal and other Mines (Electricity) Regulations 1956.	Health and Safety Commission.
(vi)	Quarry installations.	Quarries (Electricity) Regulations 1956.	Health and Safety Commission.
(vii)	Metalliferous mine installations.	Miscellaneous Mines (Electricity) Regulations 1956.	Health and Safety Commission.
(viii)	Agricultural and horticultural installations.	Agriculture (Stationary Machinery) Regulations 1959.	Health and Safety Commission.

2. Failure to comply in a consumer's installation in Great Britain with the requirements of Chapter 13 of the IEE Wiring Regulations places the supply undertaking in the position of not being compelled to commence or, in certain circumstances, to continue to give, a supply of energy to that installation.

Under Regulation 33 of the Electricity Supply Regulations 1937, any difference which may arise between a consumer and the supply undertaking having reference to the consumer's installation shall be determined, in England and Wales, by an inspector nominated by the Secretary of State for Energy and, in Scotland, by an inspector nominated by the Secretary of State, on the application of the consumer or his authorised agent or the supply undertaking.

3. It is understood that as part of the general revision of legislation under the Health and Safety at Work etc., Act it is intended to make new electricity regulations which will replace some of those referred to above. The proposed new regulations will impose requirements for electrical safety in all employment situations and will apply not only to factories but also to premises such as hospitals, teaching establishments, research premises, theatres, hotels, offices, shops and railway premises.

4. Where it is intended to use Protective Multiple Earthing the supply undertaking and the consumer must comply with the Protective Multiple Earthing Approval 1974 administered under the authority of the Secretary of State for Energy. Further information is to be found in the Explanatory Memorandum to the PME Approval 1974 issued by the Department of Energy (Electricity Division).

5. For further guidance on the application of some other of the above-mentioned statutory regulations, reference may be made to the following publications:

(i) Explanatory Notes on the Electricity Supply Regulations 1937.

(ii) Memorandum by H.M. Senior Electrical Inspector of Factories on the Electricity (Factories Act) Special Regulations (SHW 928).

(iii) Explanatory Memoranda on the Building Standards (Scotland) Amendment Regulations 1980.

6. For installations in potentially explosive atmospheres reference should be made to:

(i) The Electricity (Factories Act) Special Regulations 1908 and 1944.

(ii) The Highly Flammable Liquids and Liquified Petroleum Gases Regulations 1972.

(iii) The Petroleum (Consolidation) Act 1928.

Under the Petroleum (Consolidation) Act local authorities are empowered to grant licences in respect of premises where petroleum spirit is stored and as the authorities may attach such conditions as they think fit, the requirements may vary from one local authority to another. Guidance may be obtained from the Home Office 'Model Code of Principles of Construction and Licensing Conditions (Part I)'.

7. The Construction (General Provisions) Regulations 1961 administered by the Health and Safety Commission apply to construction sites and contain regulations concerning precautions to be taken against contact with overhead lines and underground cables encountered on construction sites. Special requirements of local authorities and of the office insuring the risk may apply to temporary installations.

8. For installations in theatres and other places of public entertainment, and on caravan sites, the requirements of the licensing authority should be ascertained. Model Standards were issued by the Department of the Environment in 1977 under the Caravan Sites and Control of Development Act 1960 as guidance for Local Authorities.

9. The Electrical Equipment (Safety) Regulations 1975 and the Electrical Equipment (Safety) (Amendment) Regulations 1976, administered by the Department of Trade, contain requirements for the safety of equipment designed or suitable for household use. Information on the application of the regulations is given in the booklet 'Administrative Guidance on the Electrical Equipment (Safety) Regulations 1975 and the Electrical Equipment (Safety) (Amendment) Regulations 1976'.

All the regulations and memoranda referred to in this Appendix, with the exception of that mentioned in Item 4, are obtainable from H.M. Stationery Office.

APPENDIX 3

EXPLANATORY NOTES ON TYPES OF SYSTEM EARTHING
(See Regulation 312–3)

In order to select the appropriate protective measures to be used for an electrical installation or the extension of an electrical installation, it is essential that certain characteristics of the source of energy for the installation are ascertained.

In particular, the choice of measures for protection against electric shock depends on, amongst other factors, the earthing arrangement at the source of energy and the type of path intended for earth fault current.

In these Regulations it is necessary to refer to 'an installation' which together with a 'source of energy' constitutes a 'system', because this is the sub-division most commonly met in practice; a public source of energy is the property and responsibility of the supply undertaking and the installation is the consumer's installation.

In some industrial situations the source of energy, the wiring and the current-using and other electrical equipment are all owned and controlled by the user and constitute what is usually called the electrical installation of the premises concerned. For the purpose of these Regulations, this combination should be considered as the electrical system, and in assessing the type of system (TN, TT, IT) the source of energy should be regarded as separate from the remainder of the equipment of the premises.

Part 2 – Definitions, includes definitions of systems using the designations 'TN', 'TT' and 'IT'.

In these designations the first letter denotes the earthing arrangement at the source of energy, and –

T = direct connection of one or more points to Earth,
I = all live parts isolated from Earth or one point connected to Earth through an impedance.

The second letter denotes the relationship of the exposed conductive parts of the installation to Earth, and

T = direct electrical connection of the exposed conductive parts to Earth, independently of the earthing of any point of the source of energy,
N = direct electrical connection of the exposed conductive parts to the earthed point of the source of energy, which, for a.c., is usually the neutral point.

NOTE – For a supply to an installation given in accordance with the Electricity Supply Regulations 1937 it may be assumed that there is direct (and permanent) connection of one or more points of the source of supply with Earth. The supply undertaking may, where practicable, provide a consumer's main earthing terminal.

The designation 'TN' is further subdivided depending on the arrangement of neutral and protective conductors, that arrangement being denoted by a further letter or letters, so that

S = neutral and protective functions provided by separate conductors
C = neutral and protective functions combined in a single conductor.

Figures 3 to 7 are explanatory schematic diagrams of examples of TN-C, TN-S, TN-C-S, TT and IT systems and are not intended to represent actual installations.

The TN-C-S system is commonly encountered where the source of energy incorporates multiple earthing of the neutral.

Approval for the use of PME is granted by the Secretary of State for Energy, or in Scotland by the Secretary of State for Scotland, to individual Electricity Boards and is subject to the conditions in the particular form of approval. Special requirements may apply to installations supplied from a PME system and the supply undertaking concerned should be consulted.

Similarly the permission of the Secretary of State for Energy, or in Scotland the Secretary of State for Scotland, is needed if a neutral derived from the neutral of a supply given in accordance with the Electricity Supply Regulations 1937 is intended to be combined with a protective conductor or be otherwise connected to Earth. Such permission, however, is not necessary if the supply is obtained from a private source.

In all cases it is the responsibility of the consumer or his agent to satisfy himself that the characteristics of the earth fault current path, including any part of that path provided by a supply undertaking, are suitable for operation of the type of earth fault protection to be used in his installation.

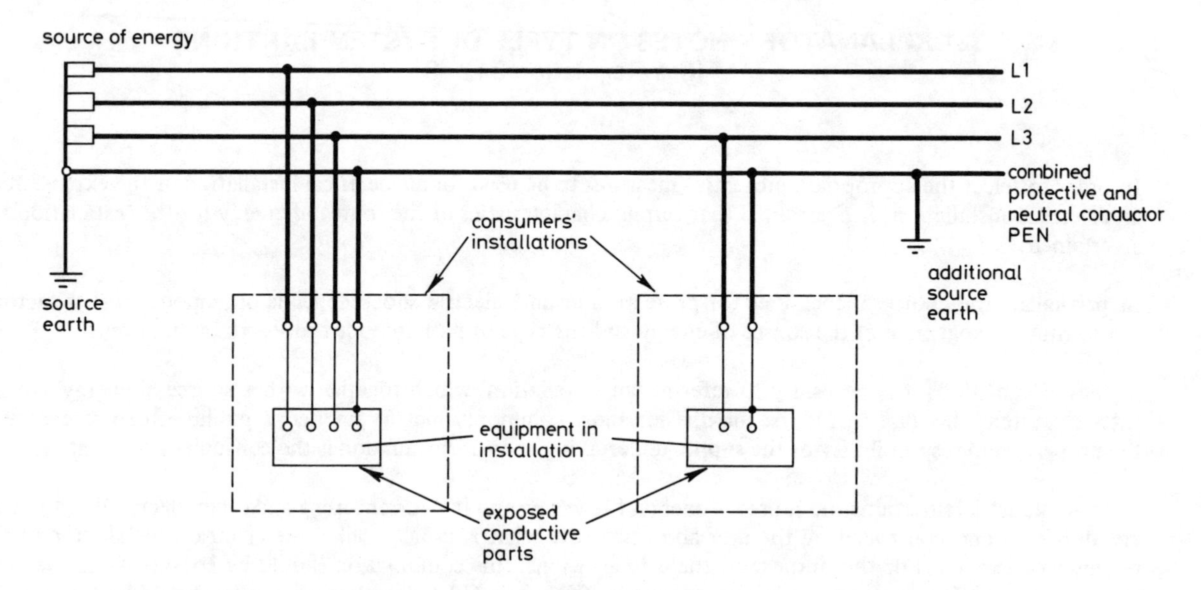

Fig. 3 TN-C system. Neutral and protective functions combined in a single conductor throughout system.

All exposed conductive parts of an installation are connected to the PEN conductor.

An example of the TN-C arrangement is earthed concentric wiring but where it is intended to use this special authorisation must be obtained from the appropriate authority.

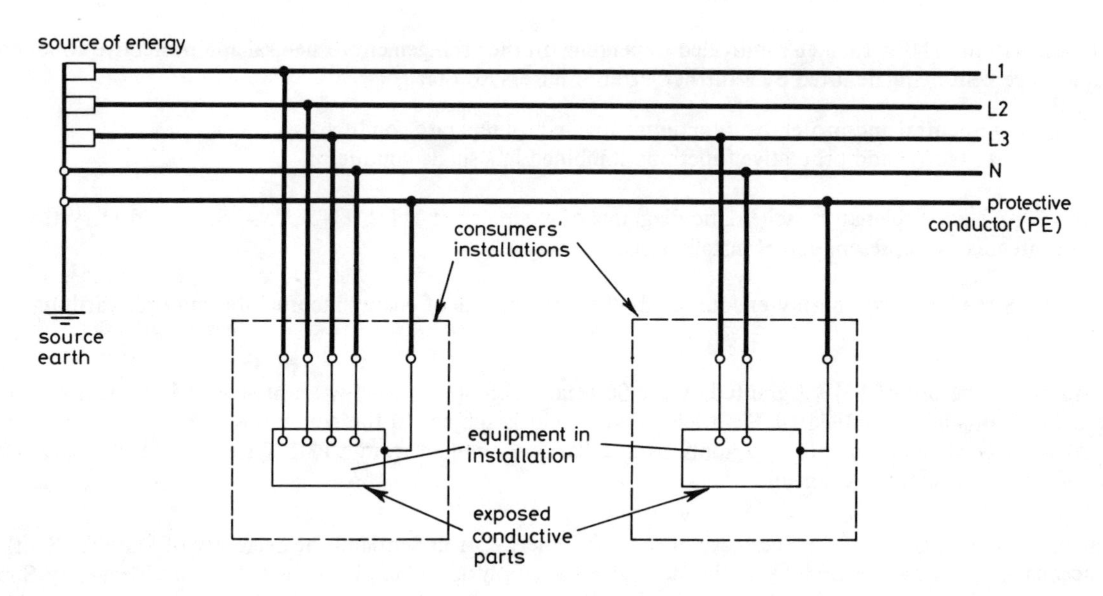

Fig. 4 TN-S system. Separate neutral and protective conductors throughout system.

The protective conductor (PE) is the metallic covering of the cable supplying the installations or a separate conductor.

All exposed conductive parts of an installation are connected to this protective conductor via the main earthing terminal of the installation.

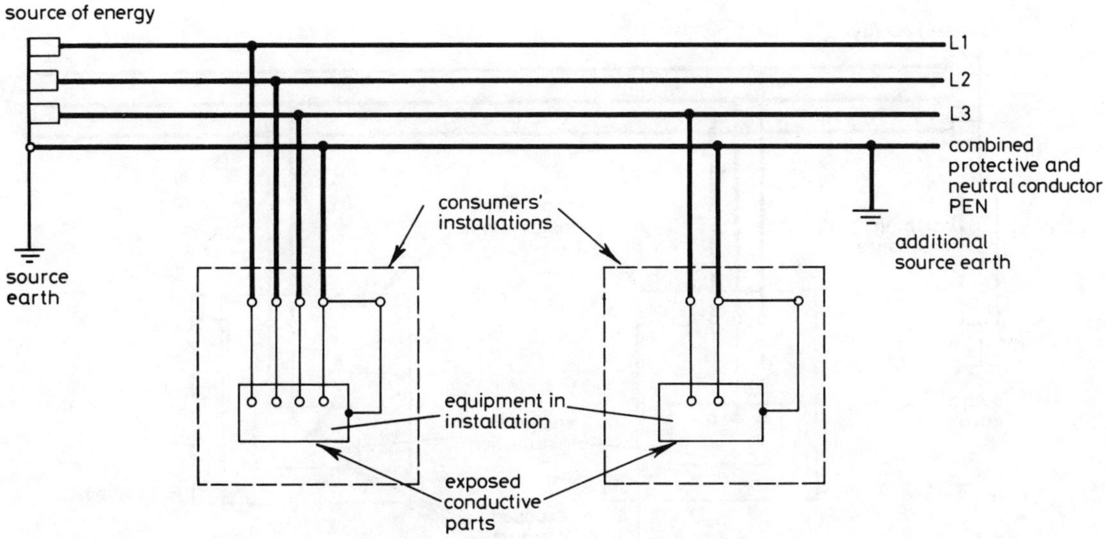

Fig. 5 **TN-C-S system. Neutral and protective functions combined in a single conductor in a part of the system.**

The usual form of a TN-C-S system is as shown, where the supply is TN-C and the arrangement in the installations is TN-S.

This type of distribution is known also as Protective Multiple Earthing and the PEN conductor is referred to as the combined neutral and earth (CNE) conductor.

The supply system PEN conductor is earthed at several points and an earth electrode may be necessary at or near a consumer's installation.

All exposed conductive parts of an installation are connected to the PEN conductor via the main earthing terminal and the neutral terminal, these terminals being linked together.

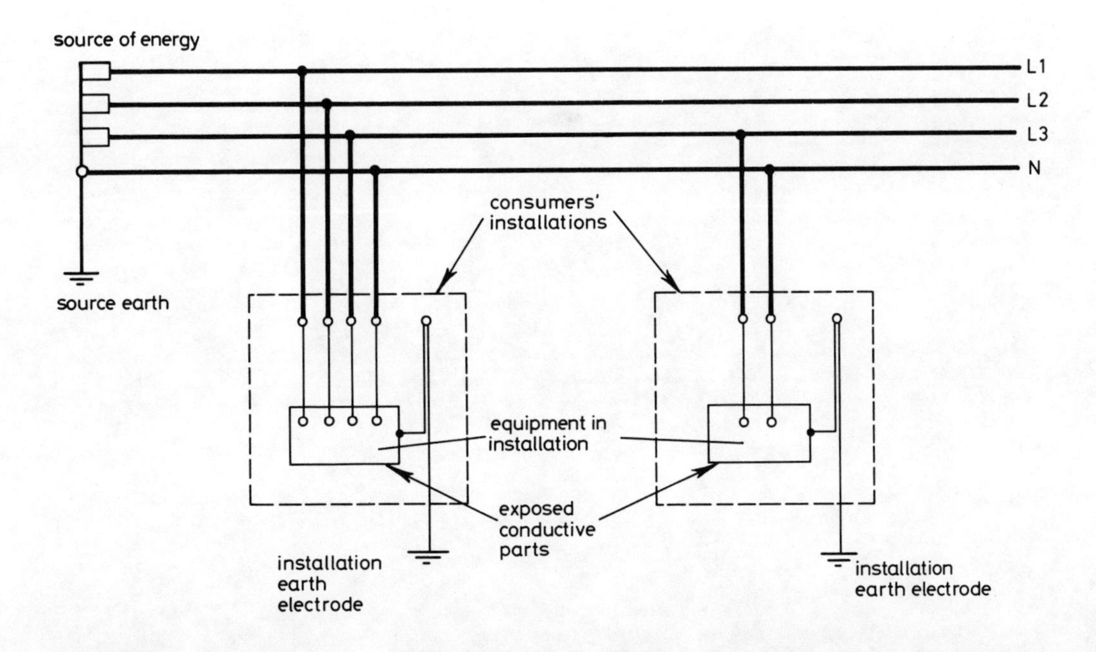

Fig. 6 **TT system**

All exposed conductive parts of an installation are connected to an earth electrode which is electrically independent of the source earth.

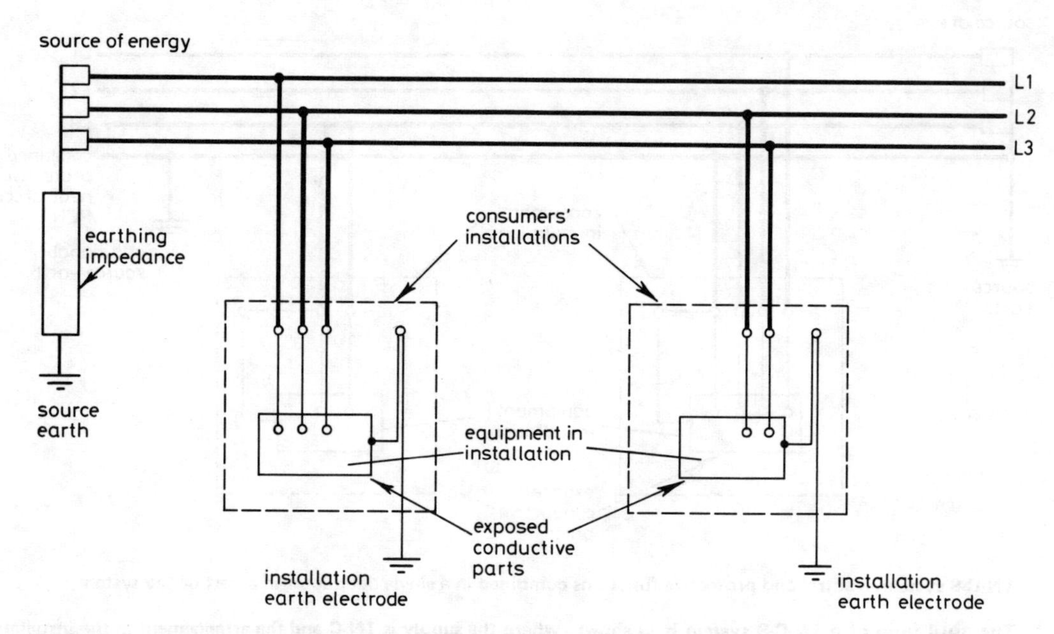

Fig. 7 IT system

All exposed conductive parts of an installation are connected to an earth electrode.

The source is either connected to Earth through a deliberately introduced earthing impedance or is isolated from Earth.

UNDER THE ELECTRICITY SUPPLY REGULATIONS 1937 THIS SYSTEM MUST NOT BE USED FOR PUBLIC SUPPLIES.

APPENDIX 4

MAXIMUM DEMAND AND DIVERSITY
(See Regulation 311–2)

This Appendix gives some information on the determination of the maximum demand for an installation and includes the current demand to be assumed for commonly used equipment. It also includes some notes on the application of allowances for diversity.

The information and values given in this Appendix are intended only for guidance because it is impossible to specify the appropriate allowances for diversity for every type of installation and such allowances call for special knowledge and experience. The figures given in Table 4B, therefore, may be increased or decreased as decided by the engineer responsible for the design of the installation concerned.

The current demand of a final circuit is determined by summating the current demands of all points of utilisation and equipment in the circuit and, where appropriate, making an allowance for diversity. Typical current demands to be used for this summation are given in Table 4A. For the standard circuits using BS 1363 socket outlets, detailed in Appendix 5, an allowance for diversity has been taken into account and no further diversity should be applied.

The current demand of a circuit supplying a number of final circuits may be assessed by using the allowances for diversity given in Table 4B which are applied to the total current demand of all the equipment supplied by that circuit and not by summating the current demands of the individual final circuits obtained as outlined above. In Table 4B the allowances are expressed either as percentages of the current demand or, where followed by the letters f.1., as percentages of the rated full load current of the current-using equipment. The current demand for any final circuit which is a standard circuit arrangement complying with Appendix 5 is the rated current of the overcurrent protective device of that circuit.

An alternative method of assessing the current demand of a circuit supplying a number of final circuits is to summate the diversified current demands of the individual circuits and then apply a further allowance for diversity but in this method the allowances given in Table 4B are not to be used, the values to be chosen being the responsibility of the designer of the installation.

The use of other methods of determining maximum demand is not precluded where specified by a suitably qualified electrical engineer.

After the design currents for all the circuits have been determined, enabling the conductor sizes to be chosen, it is necessary to check that the limitation on voltage drop is met.

TABLE 4A

Current demand to be assumed for points of utilisation and current-using equipment

Point of utilisation or current-using equipment	Current demand to be assumed
Socket outlets other than 2A socket outlets	Rated current
2A socket outlets	at least 0.5A
Lighting outlet*	Current equivalent to the connected load, with a minimum of 100W per lampholder
Electric clock, electric shaver supply unit (complying with BS 3052), shaver socket outlet (complying with BS 4573), bell transformer, and current-using equipment of a rating not greater than 5VA	May be neglected
Household cooking appliance	The first 10A of the rated current plus 30% of the remainder of the rated current plus 5A if a socket outlet is incorporated in the control unit
All other stationary equipment	British Standard rated current, or normal current.

*NOTE – Final circuits for discharge lighting are arranged so as to be capable of carrying the total steady current, viz. that of the lamp(s) and any associated gear and also their harmonic currents. Where more exact information is not available, the demand in volt-amperes is taken as the rated lamp watts multiplied by not less than 1.8. This multiplier is based upon the assumption that the circuit is corrected to a power factor of not less than 0.85 lagging, and takes into account control gear losses and harmonic currents.

TABLE 4B

Allowances for diversity

Purpose of final circuit fed from conductors or switchgear to which diversity applies	Type of premises+		
	Individual household installations, including individual dwellings of a block	Small shops, stores, offices and business premises	Small hotels, boarding houses, guest houses, etc.
1. Lighting	66% of total current demand	90% of total current demand	75% of total current demand
2. Heating and power (but see 3 to 8 below)	100% of total current demand up to 10 amperes + 50% of any current demand in excess of 10 amperes	100% f.l. of largest appliance + 75% f.l. of remaining appliances	100% f.l. of largest appliance + 80% f.l. of 2nd largest appliance +60% f.l. of remaining appliances
3. Cooking appliances	10 amperes + 30% f.l. of connected cooking appliances in excess of 10 amperes + 5 amperes if socket-outlet incorporated in unit	100% f.l. of largest appliance + 80% f.l. of 2nd largest appliance + 60% f.l. of remaining appliances	100% f.l. of largest appliance + 80% f.l. of 2nd largest appliance + 60% f.l. of remaining appliances
4. Motors (other than lift motors which are subject to special consideration)		100% f.l. of largest motor + 80% f.l. of 2nd largest motor + 60% f.l. of remaining motors	100% f.l. of largest motor + 50% f.l. of remaining motors
5. Water-heaters (instantaneous type)*	100% f.l. of largest appliance + 100% f.l. of 2nd largest appliance + 25% f.l. of remaining appliances	100% f.l. of largest appliance + 100% f.l. of 2nd largest appliance + 25% f.l. of remaining appliances	100% f.l. of largest appliance + 100% f.l. of 2nd largest appliance + 25% f.l. of remaining appliances
6. Water-heaters (thermostatically controlled)	no diversity allowable†		
7. Floor warming installations	no diversity allowable†		
8. Thermal storage space heating installations	no diversity allowable†		
9. Standard arrangements of final circuits in accordance with Appendix 5	100% of current demand of largest circuit + 40% of current demand of every other circuit	100% of current demand of largest circuit + 50% of current demand of every other circuit	
10. Socket outlets other than those included in 9 above and stationary equipment other than those listed above	100% of current demand of largest point of utilisation + 40% of current demand of every other point of utilisation	100% of current demand of largest point of utilisation + 75% of current demand of every other point of utilisation	100% of current demand of largest point of utilisation + 75% of current demand of every point in main rooms (dining rooms, etc) + 40% of current demand of every other point of utilisation

+ For blocks of residential dwellings, large hotels, large commercial premises, and factories, the allowances are to be assessed by a competent person.

* For the purpose of this Table an instantaneous water-heater is deemed to be a water-heater of any loading which heats water only while the tap is turned on and therefore uses electricity intermittently.

† It is important to ensure that the distribution boards are of sufficient rating to take the total load connected to them without the application of any diversity.

APPENDIX 5

STANDARD CIRCUIT ARRANGEMENTS

This appendix gives details of standard circuit arrangements which, as stated in Regulation 314–3, satisfy the requirements of Chapter 43 for overcurrent protection and Chapter 46 for isolation and switching, together with the requirements as regards current-carrying capacities of conductors prescribed in Chapter 52 – Cables, conductors and wiring materials.

It is the responsibility of the designer and installer when adopting these circuit arrangements to take the appropriate measures to comply with the requirements of other chapters or sections which are relevant, such as Chapter 41 – Protection against electric shock, Chapter 54 – Earthing and protective conductors, and the requirements of Chapter 52 other than those concerning current-carrying capacities.

Circuit arrangements other than those detailed in this appendix are not precluded where they are specified by a suitably qualified electrical engineer, in accordance with the general requirements of Regulation 314–3.

The standard circuit arrangements are:

- Final circuits using socket outlets complying with BS 1363
- Final circuits using socket outlets complying with BS 196
- Final radial circuits using socket outlets complying with BS 4343
- Cooker final circuits in household premises.

Final circuits using socket outlets complying with BS 1363 and fused connection units

General

A ring or radial circuit, with spurs if any, feeds permanently connected equipment and an unlimited number of socket outlets.

The floor area served by the circuit is determined by the known or estimated load but does not exceed the value given in Table 5A.

For household installations a single 30A ring circuit may serve a floor area of up to 100m² but consideration should be given to the loading in kitchens which may require a separate circuit. For other types of premises, final circuits complying with Table 5A may be installed where, owing to diversity, the maximum demand of current-using equipment to be connected is estimated not to exceed the corresponding ratings of the overcurrent protective device given in that table.

The number of socket outlets is such as to ensure compliance with Regulation 553–9, each socket outlet of a twin or multiple socket outlet unit being regarded as one socket outlet.

Diversity between socket outlets and permanently connected equipment has already been taken into account in Table 5A and no further diversity should be applied.

TABLE 5A

Final circuits using BS 1363 socket outlets

Type of circuit	Overcurrent protective device		Minimum conductor size*			
			Copper conductor rubber- or p.v.c.-insulated cables	Copperclad aluminium conductor p.v.c.-insulated cables	Copper conductor mineral-insulated cables	Maximum floor area served
	Rating	Type				
1	2	3	4	5	6	7
	A		mm²	mm²	mm²	m²
A1 Ring	30 or 32	Any	2.5	4	1.5	100
A2 Radial	30 or 32	Cartridge fuse or circuit breaker	4	6	2.5	50
A3 Radial	20	Any	2.5	4	1.5	20

*The tabulated values of conductor size may be reduced for fused spurs.

Immersion heaters, fitted to storage vessels in excess of 15 litres capacity, or permanently connected heating appliances forming part of a comprehensive space heating installation are supplied by their own separate circuits.

Where two or more ring final circuits are installed the socket outlets and permanently connected equipment to be served are reasonably distributed among the circuits.

Circuit protection

The overcurrent protective device is of the type, and has the rating, given in Table 5A.

Conductor size

The minimum size of conductor in the circuit and in non-fused spurs is given in Table 5A. However, if the cables of more than two circuits are bunched together or the ambient temperature exceeds 30°C, the size of conductor is increased and is determined by applying the appropriate correction factors from Appendix 9 such that the size then corresponds to a current-carrying capacity of –

20A for circuit A1 (i.e. 0.67 times the rating of the overcurrent protective device),

30A or 32A for circuit A2 (i.e. the rating of the overcurrent protective device),

20A for circuit A3 (i.e. the rating of the overcurrent protective device).

The conductor size for a fused spur is determined from the total current demand served by that spur, which is limited to a maximum of 13A.

When such a spur serves socket outlets the minimum conductor size is:

1.5mm^2 for rubber- or p.v.c.-insulated cables, copper conductors,

2.5mm^2 for rubber- or p.v.c.-insulated cables, copperclad aluminium conductors,

1mm^2 for mineral-insulated cables, copper conductors.

Spurs

The total number of fused spurs is unlimited but the number of non-fused spurs does not exceed the total number of socket outlets and items of stationary equipment connected directly in the circuit.

A non-fused spur feeds only one single or one twin socket outlet or one permanently connected equipment. Such a spur is connected to a circuit at the terminals of socket outlets or at joint boxes or at the origin of the circuit in the distribution board.

A fused spur is connected to the circuit through a fused connection unit, the rating of the fuse in the unit not exceeding that of the cable forming the spur, and, in any event, not exceeding 13A.

Permanently connected equipment

Permanently connected equipment is locally protected by a fuse of rating not exceeding 13A and controlled by a switch conforming with the requirements of Regulation 476–17, or protected by a circuit breaker of rating not exceeding 16A.

Final circuits using socket outlets complying with BS 196

General

A ring or radial circuit, with fused spurs if any, feeds equipment the maximum demand of which, having allowed for diversity, is known or estimated not to exceed the rating of the overcurrent protective device and in any event does not exceed 32A.

In assessing the maximum demand it is assumed that permanently connected equipment operates continuously, i.e., no diversity is allowed in respect of such equipment.

The number of socket outlets is unlimited.

The total current demand of points served by a fused spur does not exceed 16A.

Circuit protection

The overcurrent protective device has a rating not exceeding 32A.

Conductor size

The size of conductor is determined by applying from Appendix 9 the appropriate correction factors and is such that it then corresponds to a current-carrying capacity of –

– for ring circuits – not less than 0.67 times the rating of the overcurrent protective device,

– for radial circuits – not less than the rating of the overcurrent protective device.

The conductor size for a fused spur is determined from the total demand served by that spur which is limited to a maximum of 16A.

Spurs

A fused spur is connected to a circuit through a fused connection unit, the rating of the fuse in the unit not exceeding that of the cable forming the spur and in any event not exceeding 16A.

Non-fused spurs are not used.

Permanently connected equipment

Permanently connected equipment is locally protected by a fuse of rating not exceeding 16A and controlled by a switch conforming with the requirements of Regulation 476–17, or by a circuit-breaker of rating not exceeding 16A.

Types of socket-outlet

If the circuit has one pole earthed the socket outlet is of the type that will accept only 2-pole-and-earth contact plugs with single-pole fusing on the live pole. Such socket outlets are those which have raised socket keys to prevent insertion of non-fused plugs, together with socket keyways recessed at position 'B' and such other positions as are specified in the British Standard according to the nature of the supply to the socket outlets.

If the circuit has neither pole earthed (e.g. a circuit supplied from a double-wound transformer having the mid-point of its secondary winding earthed) the socket outlet is of the type that will accept only 2-pole-and-earth contact plugs with double-pole fusing. Such socket outlets are those which have raised socket keys to prevent insertion of non-fused plugs, together with socket keyways recessed at position 'P' and such other positions as are specified in the British Standard according to the nature of the supply to the socket outlets.

Final radial circuits using 16A socket outlets complying with BS 4343

General

A radial circuit feeds equipment the maximum demand of which, having allowed for diversity, is known or estimated not to exceed the rating of the overcurrent protective device and in any event does not exceed 20A.

The number of socket outlets is unlimited.

Circuit protection

The overcurrent protective device has a rating not exceeding 20A.

Conductor size

The size of conductor is determined by applying from Appendix 9 the appropriate correction factors and is such that it then corresponds to a current-carrying capacity not less than the rating of the overcurrent protective device.

Types of socket outlets

Socket outlets have a rated current of 16A and are of the type appropriate to the number of phases, circuit voltage and earthing arrangement. Socket outlets incorporating pilot contacts are not included.

Cooker circuits in household premises

The circuit supplies a control switch or a cooker control unit which may incorporate a socket outlet.

The rating of the circuit is determined by the assessment of the current demand of the cooking appliance(s), and control unit socket outlet if any, in accordance with Table 4A of Appendix 4.

A circuit of rating exceeding 15A but not exceeding 50A may supply two or more cooking appliances where these are installed in one room. (See also Regulation 476–20).

Attention is drawn to the need to afford discriminative operation of protective gear as stated in Regulation 533–6.

APPENDIX 6

CLASSIFICATION OF EXTERNAL INFLUENCES
(See note to Chapter 32)

This appendix gives the classification and codification of external influences developed for IEC Publication 364.

Each condition of external influences is designated by a code comprising a group of two capital letters and a number, as follows:

The first letter relates to the general category of external influence:

A. Environment

B. Utilisation

C. Construction of buildings

The second letter relates to the nature of the external influence:

A. ...

B. ...

C. ...

The number relates to the class within each external influence:

1. ...

2. ...

3. ...

For example the code AA4 signifies:

A = Environment

AA = Environment – Ambient temperature

AA4 = Environment – Ambient temperature in the range of −5°C to 40°C.

NOTE – The codification given in this appendix is not intended to be used for marking equipment.

TABLE 6A

Classes of external influence – environmental conditions

Code	Class Designation	Characteristics	Applications and Examples
		Ambient temperature	
		The ambient temperature is that of the ambient air where the equipment is to be installed.	
		It is assumed that the ambient temperature includes the effects of all other equipment installed in the same location.	
		The ambient temperature to be considered for the equipment is the temperature at the place where the equipment is to be installed resulting from the influence of all other equipment in the same location, when operating, not taking into account the thermal contribution of the equipment to be installed.	
		Lower and upper limits of the ranges of ambient temperature:	
AA1		-60°C + 5°C	
AA2		-40°C + 5°C	
AA3		-25°C + 5°C	
AA4		- 5°C +40°C	
AA5		+ 5°C +40°C	
AA6		+ 5°C +60°C	
		The average temperature over a 24-hour period must not exceed 5°C below the upper limits.	
		Combination of two ranges to define some environments may be necessary. Installations subject to temperatures outside the ranges require special consideration.	
		Atmospheric humidity	
AB	(Under consideration)		
		Altitude	
AC1		⩽2000 m	
AC2		>2000 m	
		Presence of water	
AD1	Negligible	Probability of presence of water is negligible.	Locations in which the walls do not generally show traces of water but may do so for short periods, for example in the form of vapour which good ventilation dries rapidly.
AD2	Free-falling drops	Possibility of vertically falling drops	Locations in which water vapour occasionally condenses as drops or where steam may occasionally be present.
AD3	Sprays	Possibility of water falling as a spray at an angle up to 60° from the vertical.	Locations in which sprayed water forms a continuous film on floors and/or walls.

TABLE 6A (contd.)

Code	Class Designation	Characteristics	Applications and Examples
AD4	Splashes	Possibility of splashes from any direction.	Locations where equipment may be subjected to splashed water; this applies, for example, to certain external lighting fittings, construction site equipment.
AD5	Jets	Possibility of jets of water from any direction.	Locations where hosewater is used regularly (yards, car-washing bays).
AD6	Waves	Possibility of water waves.	Seashore locations such as piers, beaches, quays, etc.
AD7	Immersion	Possibility of intermittent partial or total covering by water.	Locations which may be flooded and/or where water may be at least 150 mm above the highest point of equipment, the lowest part of equipment being not more than 1 m below the water surface.
AD8	Submersion	Possibility of permanent and total covering by water.	Locations such as swimming pools where electrical equipment is permanently and totally covered with water under a pressure greater than 0.1 bar.
		Presence of foreign solid bodies	
AE1	Negligible	The quantity or nature of dust or foreign solid bodies is not significant.	
AE2	Small objects	Presence of foreign solid bodies where the smallest dimension is not less than 2.5 mm.	Tools and small objects are examples of foreign solid bodies of which the smallest dimension is at least 2.5 mm.
AE3	Very small objects	Presence of foreign solid bodies where the smallest dimension is not less than 1 mm.	Wires are examples of foreign solid bodies of which the smallest dimension is not less than 1 mm.
		Note: In conditions AE2 and AE3, dust may be present but is not significant to operation of the electrical equipment.	
AE4	Dust	Presence of dust in significant quantity.	
		Presence of corrosive or polluting substances	
AF1	Negligible	The quantity or nature of corrosive or polluting substances is not significant.	
AF2	Atmospheric	The presence of corrosive or polluting substances of atmospheric origin is significant.	Installations situated by the sea or industrial zones producing serious atmospheric pollution, such as chemical works, cement works; this type of pollution arises especially in the production of abrasive, insulating or conductive dusts.

TABLE 6A (contd.)

Code	Class Designation	Characteristics	Applications and Examples
AF3	Intermittent or accidental	Intermittent or accidental subjection to corrosive or polluting chemical substances being used or produced.	Locations where some chemical products are handled in small quantities and where these products may come only accidentally into contact with electrical equipment; such conditions are found in factory laboratories, other laboratories, or in locations where hydrocarbons are used (boiler rooms, garages, etc.).
AF4	Continuous	Continuously subject to corrosive or polluting chemical substances in substantial quantity.	For example, chemical works.
	Mechanical stresses *Impact*		
AG1	Low severity	Note: Provisional classification. Quantitative expression of impact severities is under consideration.	Household and similar conditions.
AG2	Medium severity		Usual industrial conditions.
AG3	High severity		Severe industrial conditions.
	Vibration		
AH1	Low severity	Note: Provisional classification. Quantitative expression of vibration severities is under consideration.	Household and similar conditions where the effects of vibration are generally negligible.
AH2	Medium severity		Usual industrial conditions.
AH3	High severity		Industrial installations subject to severe conditions.
	Other mechanical stresses		
AJ	(Under consideration)		
	Presence of flora and/or mould growth		
AK1	No hazard	No harmful hazard of flora and/or mould growth.	
AK2	Hazard	Harmful hazard of flora and/or mould growth.	The hazard depends on local conditions and the nature of flora. Distinction should be made between harmful growth of vegetation or conditions for promotion of mould growth.
	Presence of fauna		
AL1	No hazard	No harmful hazard from fauna.	The hazard depends on the nature of the fauna. Distinction should be made between: – presence of insects in harmful quantity or of an aggressive nature – presence of small animals or birds in harmful quantity or of an aggressive nature.
AL2	Hazard	Harmful hazard from fauna. (insects, birds, small animals)	

TABLE 6A (contd.)

Code	Class Designation	Characteristics	Applications and Examples
		Electromagnetic, electrostatic or ionizing influences	
AM1	Negligible	No harmful effects from stray currents, electromagnetic radiation, electrostatic fields, ionizing radiation or induction.	
AM2	Stray currents	Harmful hazards of stray currents.	
AM3	Electromagnetics	Harmful presence of electromagnetic radiation.	
AM4	Ionization	Harmful presence of ionizing radiation.	
AM5	Electrostatics	Harmful presence of electrostatic fields.	
AM6	Induction	Harmful presence of induced currents.	
		Solar radiation	
AN1	Negligible	–	
AN2	Significant	Solar radiation of harmful intensity and/or duration.	
		Seismic effects	
AP1	Negligible	⩽ 30 Gal	1 Gal = 1 cm/s^2
AP2	Low severity	30 < Gal ⩽ 300	
AP3	Medium severity	300 < Gal ⩽ 600	Vibration which may cause the destruction of the building is outside the classification.
AP4	High severity	> 600 Gal	Frequency is not taken into account in the classification; however, if the seismic wave resonates with the building, seismic effects must be specially considered. In general, the frequency of seismic acceleration is between 0 and 10 Hz.
		Lightning	
AQ1	Negligible	–	
AQ2	Indirect exposure	Hazard from supply arrangements.	Installations supplied by overhead lines.
AQ3	Direct exposure	Hazard from exposure of equipment.	Parts of installations located outside buildings. The risks AQ2 and AQ3 relate to regions with a particularly high level of thunderstorm activity.
		Wind	
AR–	(Under consideration)		

TABLE 6B
Classes of external influence – Utilisation

Code	Class Designation	Characteristics	Applications and Examples
		Capability of persons	
BA1	Ordinary	Uninstructed persons.	
BA2	Children	Children in locations intended for their occupation. Note: This class does not necessarily apply to family dwellings.	Nurseries.
BA3	Handicapped	Persons not in command of all their physical and intellectual abilities (sick persons, old persons).	Hospitals.
BA4	Instructed	Persons adequately advised or supervised by skilled persons to enable them to avoid dangers which electricity may create (operating and maintenance staff).	Electrical operating areas.
BA5	Skilled	Persons with technical knowledge or sufficient experience to enable them to avoid dangers which electricity may create (engineers and technicians).	Closed electrical operating areas.
		Electrical resistance of the human body	
BB	Classification under consideration		
		Contact of persons with earth potential	
BC1	None	Persons in non-conducting situations.	Non-conducting locations.
BC2	Low	Persons do not in usual conditions make contact with extraneous conductive parts or stand on conducting surfaces.	
BC3	Frequent	Persons are frequently in touch with extraneous conductive parts or stand on conducting surfaces.	Locations with extraneous conductive parts, either numerous or of large area.
BC4	Continuous	Persons are in permanent contact with metallic surroundings and for whom the possibility of interrupting contact is limited.	Metallic surroundings such as boilers and tanks.
		Conditions of evacuation in an emergency	
BD1		Low density occupation, easy conditions of evacuation.	Buildings of normal or low height used for habitation.
BD2		Low density occupation, difficult conditions of evacuation.	High-rise buildings.
BD3		High density occupation, easy conditions of evacuation.	Locations open to the public (theatres, cinemas, department stores, etc.).
BD4		High density occupation, difficult conditions of evacuation.	High-rise buildings open to the public (hotels, hospitals, etc.).
		Nature of processed or stored materials	
BE1	No significant risks	–	–
BE2	Fire risks	Manufacture, processing or storage of flammable materials including presence of dust.	Barns, wood-working shops, paper factories.

TABLE 6B (contd.)

Code	Class Designation	Characteristics	Applications and Examples
BE3	Explosion risks	Processing or storage of explosive or low-flashpoint materials including presence of explosive dusts.	Oil refineries, hydrocarbon stores.
BE4	Contamination risks	Presence of unprotected foodstuffs, pharmaceutics, and similar products without protection.	Foodstuff industries, kitchens.

TABLE 6C

Classes of external influence – Construction of buildings

Code	Class Designation	Characteristics	Applications and Examples
	Constructional materials		
CA1	Non-combustible	–	–
CA2	Combustible	Buildings mainly constructed of combustible materials.	Wooden buildings
	Building design		
CB1	Negligible risks	–	–
CB2	Propagation of fire	Buildings of which the shape and dimensions facilitate the spread of fire (e.g. chimney effects).	High-rise buildings. Forced ventilation systems.
CB3	Movement	Risks due to structural movement (e.g. displacement between different parts of a building or between a building and the ground, or settlement of ground or building foundations).	Buildings of considerable length or erected on unstable ground.
CB4	Flexible or unstable.	Structures which are weak or subject to movement (e.g. oscillation).	Tents, air-support structures, false ceilings, removable partitions.

APPENDIX 7

AN ALTERNATIVE METHOD OF COMPLIANCE WITH REGULATION 413–3 FOR CIRCUITS SUPPLYING SOCKET OUTLETS

Introduction

Regulation 413–4 specifies one method of complying with Regulation 413–3 by placing a limitation on the earth fault loop impedance of a circuit so that in the event of an earth fault the time for disconnection for a circuit supplying socket outlets does not exceed 0.4 second. Table 41A 1 of Regulation 413–5 gives the limiting values of earth fault loop impedance appropriate to various types and ratings of overcurrent protective device so that this disconnection time is not exceeded, assuming that the fault is of negligible impedance.

As indicated in item (i) of Regulation 413–4 other methods of complying with Regulation 413–3 are not precluded and the purpose of this appendix is to detail one such method which, like that given in Regulations 413–4 and 413–5, is applicable only where the exposed conductive parts of the equipment concerned and any extraneous conductive parts are situated within the zone created by the main equipotential bonding (see Regulation 471–12).

The alternative method

This alternative method involves limiting the impedance of the protective conductor of the circuit concerned to the main earthing terminal (Z_2) and permits the disconnection time in the event of an earth fault to be increased from 0.4 second to 5 seconds (the latter being the same as for a circuit supplying fixed equipment).

Two tables of limiting values of protective conductor impedance are given for various types and ratings of overcurrent protective devices, as follows:

– Table 7A applies when the protective conductor is not in the same cable as the associated phase conductors, and

– Table 7B applies when the protective conductor is in the same cable as the associated phase conductors.

Both tables are based on a conductor temperature of 30°C and, except for the values given for semi-enclosed fuses to BS 3036, apply for all values of nominal voltage to earth (U_o). The values given for semi-enclosed fuses apply for values of U_o not exceeding 240V.

Application of method to a radial circuit

The impedance of the protective conductor from the most distant socket outlet or point of utilisation of a radial circuit to the main earthing terminal shall not exceed the appropriate value given in Table 7A or Table 7B. In addition, the earth fault loop impedance shall not exceed the appropriate value given in Table 41A 2 or that required for compliance with Regulation 543–1 whichever is the smaller.

Application of method to a ring circuit

The impedance of the protective conductor measured between its two ends prior to completion of the ring shall not exceed four times the appropriate value given in Table 7A or Table 7B. In addition, for a spurred ring circuit the appropriate tabulated value shall not be exceeded at any point of utilisation on a spur.

Included in both Tables are the relevant values for 13A fuses to BS 1362 for use when the spur is fed from a fused connection box.

In addition, the earth fault loop impedance shall not exceed the appropriate value given in Table 41A 2 or that required for compliance with Regulation 543–1 whichever is the smaller.

TABLE 7A

Maximum impedance of protective conductor when not in the same cable as associated phase conductors

(a) Fuses to BS 88 Part 2

Rating Amperes	6	10	16	20	25	32	40	50
Z_2 ohms	2.78	1.61	0.92	0.65	0.5	0.39	0.29	0.24

(b) Fuses to BS 1361

Rating Amperes	5	15	20	30	45
Z_2 ohms	3.57	1.08	0.63	0.42	0.21

(c) Fuses to BS 3036

Rating Amperes	5	15	20	30	45
Z_2 ohms	2.0	0.55	0.38	0.24	0.125

(d) Fuse to BS 1362

Rating Amperes	13
Z_2 ohms	0.83

(e) Type 1 miniature circuit breakers to BS 3871

Rating Amperes	5	10	15	20	30	50	I_n
Z_2 ohms	2.5	1.25	0.83	0.63	0.41	0.25	$\frac{12.5}{I_n}$

(f) Type 2 miniature circuit breakers to BS 3871

Rating Amperes	5	10	15	20	30	50	I_n
Z_2 ohms	1.42	0.71	0.47	0.35	0.24	0.14	$\frac{7.1}{I_n}$

(g) Type 3 miniature circuit breakers to BS 3871

Rating Amperes	5	10	15	20	30	50	I_n
Z_2 ohms	1.06	0.53	0.35	0.26	0.17	0.10	$\frac{5.3}{I_n}$

TABLE 7B

Maximum impedance of protective conductor when in the same cable as associated phase conductors

(a) Fuses to BS 88 Part 2

Rating Amperes	6	10	16	20	25	32	40	50
Z_2 ohms	2.4	1.38	0.79	0.56	0.43	0.33	0.24	0.21

(b) Fuses to BS 1361

Rating Amperes	5	15	20	30	45
Z_2 ohms	3.03	0.92	0.53	0.35	0.18

(c) Fuses to BS 3036

Rating Amperes	5	15	20	30	45
Z_2 ohms	1.72	0.47	0.32	0.21	0.11

(d) Fuse to BS 1362

Rating Amperes	13
Z_2 ohms	0.7

(e) Type 1 miniature circuit breakers to BS 3871

Rating Amperes	5	10	15	20	30	50	I_n
Z_2 ohms	2.12	1.06	0.7	0.53	0.35	0.21	$\frac{10.6}{I_n}$

(f) Type 2 miniature circuit breakers to BS 3871

Rating Amperes	5	10	15	20	30	50	I_n
Z_2 ohms	1.2	0.6	0.4	0.3	0.2	0.12	$\frac{6}{I_n}$

(g) Type 3 miniature circuit breakers to BS 3871

Rating Amperes	5	10	15	20	30	50	I_n
Z_2 ohms	0.9	0.45	0.3	0.22	0.14	0.08	$\frac{4.5}{I_n}$

APPENDIX 8

LIMITATION OF EARTH FAULT LOOP IMPEDANCE FOR COMPLIANCE WITH REGULATION 543–1

Regulation 543–1 indicates that the cross-sectional area of a protective conductor, other than an equipotential bonding conductor, shall be –

(i) calculated in accordance with Regulation 543–2, or

(ii) selected in accordance with Regulation 543–3.

In some cases the type of cable it is intended to use determines which of the two methods can be followed. For instance, the widely used flat twin and flat three-core p.v.c.-insulated and p.v.c.-sheathed cables with protective conductors (cables to Table 5 of BS 6004) do not comply with Table 54F of Regulation 543–3 (other than the 1mm^2 size) and therefore Method (i) must be used.

Where Method (i) is used, in order to apply the formula given in Regulation 543–2 it is essential that the time/current characteristic of the overcurrent protective device in the circuit concerned is available. A selection of such characteristics for fuses and miniature circuit breakers is given in this appendix but for other types of these devices the advice of the manufacturer has to be sought. The time/current characteristics given in this appendix indicate the maximum disconnection times for the devices concerned.

Assuming that the size and type of cable to be used has already been determined from consideration of other aspects such as the magnitude of the design current of the circuit and the limitation of voltage drop under normal load conditions, the first stage is to calculate the earth fault loop impedance, Z_s. If the cable it is intended to use does not incorporate a protective conductor, that conductor has to be chosen separately.

For cables having conductors of cross-sectional area less than 35mm^2, their inductance can be ignored so that where these cables are used in radial circuits, the earth fault loop impedance Z_s is given by –

$$Z_s = Z_E + R_1 + R_2 \text{ ohms}$$

Where Z_E is that part of the earth fault loop impedance external to the circuit concerned, R_1 is the resistance of the phase conductor from the origin of the circuit to the most distant socket outlet or other point of utilisation, and R_2 is the resistance of the protective conductor from the origin of the circuit to the most distant socket outlet or other point of utilisation.

Similarly, where such cables are used in ring circuits without spurs, the earth fault loop impedance Z_s is given by–

$$Z_s = Z_E + 0.25\,R_1 + 0.25\,R_2 \text{ ohms}$$

Where Z_E is as described above, R_1 is now the total resistance of the phase conductor between its ends prior to them being connected together to complete the ring, and R_2 is similarly the total resistance of of the protective conductor.

NOTE – Strictly the above equations are vectorial but arithmetic addition to determine the earth fault loop impedance may be used, as it gives a pessimistically high value for that impedance.

Having determined Z_s, the earth fault current I_F, is given by –

$$I_F = \frac{U_o}{Z_s} \text{ amperes}$$

Where U_o is the nominal voltage to Earth (phase to neutral voltage).

From the relevant time/current characteristic the time for disconnection (t) corresponding to this earth fault current is obtained.

Substitution for I_F, t and the appropriate k value in the equation given in Regulation 543–2 then gives the minimum cross-sectional area of protective conductor and this has to be equal to or less than the size chosen.

In order to assist the designer, when the cables it is intended to use are to Table 5 of BS 6004, or are other p.v.c.-insulated cables to that British Standard, Tables 8A to 8C give the maximum earth loop impedances for circuits having protective conductors of copper and from 1mm^2 to 16mm^2 cross-sectional area and where the overcurrent protective device is either a fuse to BS 88 Part 2, BS 1361 or BS 3036 or is a miniature circuit breaker to BS 3871. The tables also apply if the protective conductor is bare copper and in contact with cable insulated with p.v.c.

For each type of fuse, two tables are given:

– where the circuit concerned feeds socket outlets and the disconnection time for compliance with Regulation 413–4 is 0.4 second, and

– where the circuit concerned feeds fixed equipment and the disconnection time for compliance with Regulation 413–4 is 5 seconds.

In each table the earth fault loop impedances given correspond to the appropriate disconnection time or a shorter time if the latter has been found necessary from a comparison of the time/current characteristic of the device concerned and the equation given in Regulation 543–2.

If in the Tables no value of earth fault loop impedance is given for a particular combination of protective conductor cross-sectional area and current rating of the fuse that combination will not meet the requirements of Chapter 54.

The tabulated values apply only when the nominal voltage to Earth (U_o) is 240 Volts.

For circuits protected by miniature circuit breakers compliance with Table 41A1 affords compliance with Regulation 543–1 where the protective conductors range from $1mm^2$ to $16mm^2$ cross-sectional area and the rated current of the miniature circuit breakers ranges from 5A to 50A.

Table 8D gives values of $(R_1 + R_2)$ per metre for copper conductors and these values apply where the protective conductor is –

– insulated with p.v.c. but not incorporated in the same cable as the associated phase conductors, or

– bare and in contact with p.v.c. covering of cables, or

– insulated or bare and incorporated in the same cable as the associated phase conductors.

Provided that Z_E is known, Table 8D enables the designer to determine the maximum length of circuit for compliance with Regulation 543–1.

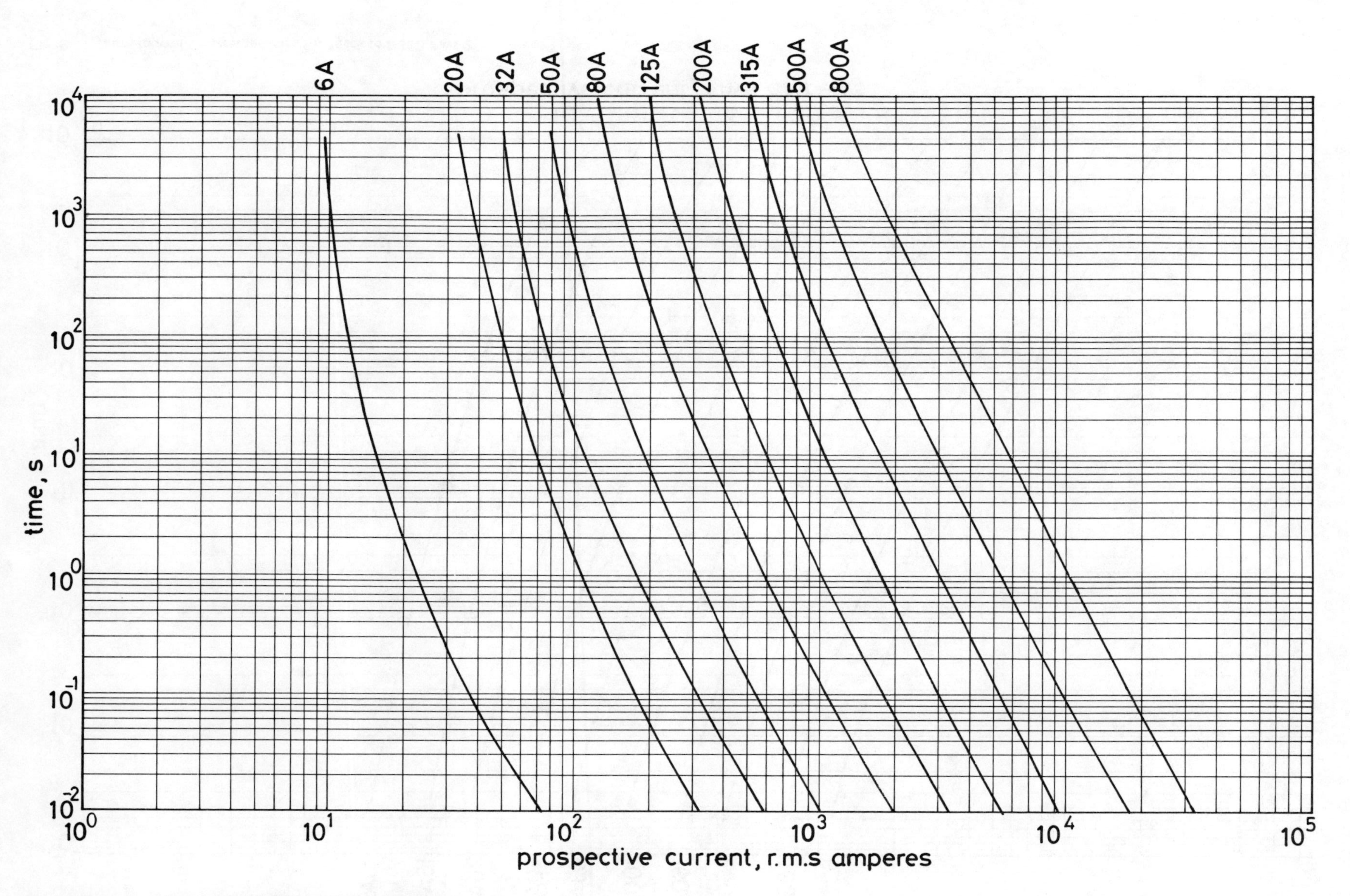

Fig. 8 Time/current characteristics for fuses to BS88 Part 2

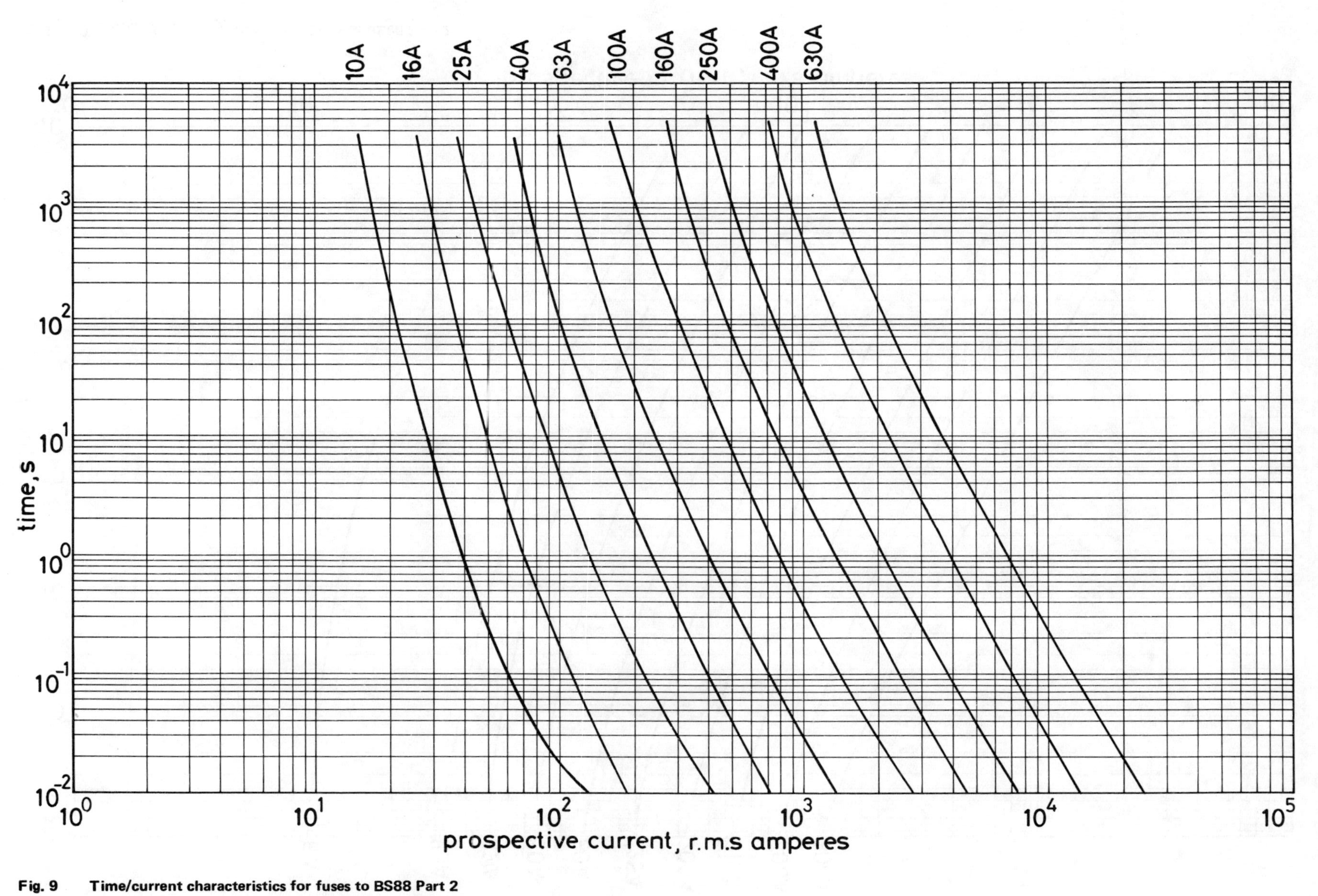

Fig. 9 Time/current characteristics for fuses to BS88 Part 2

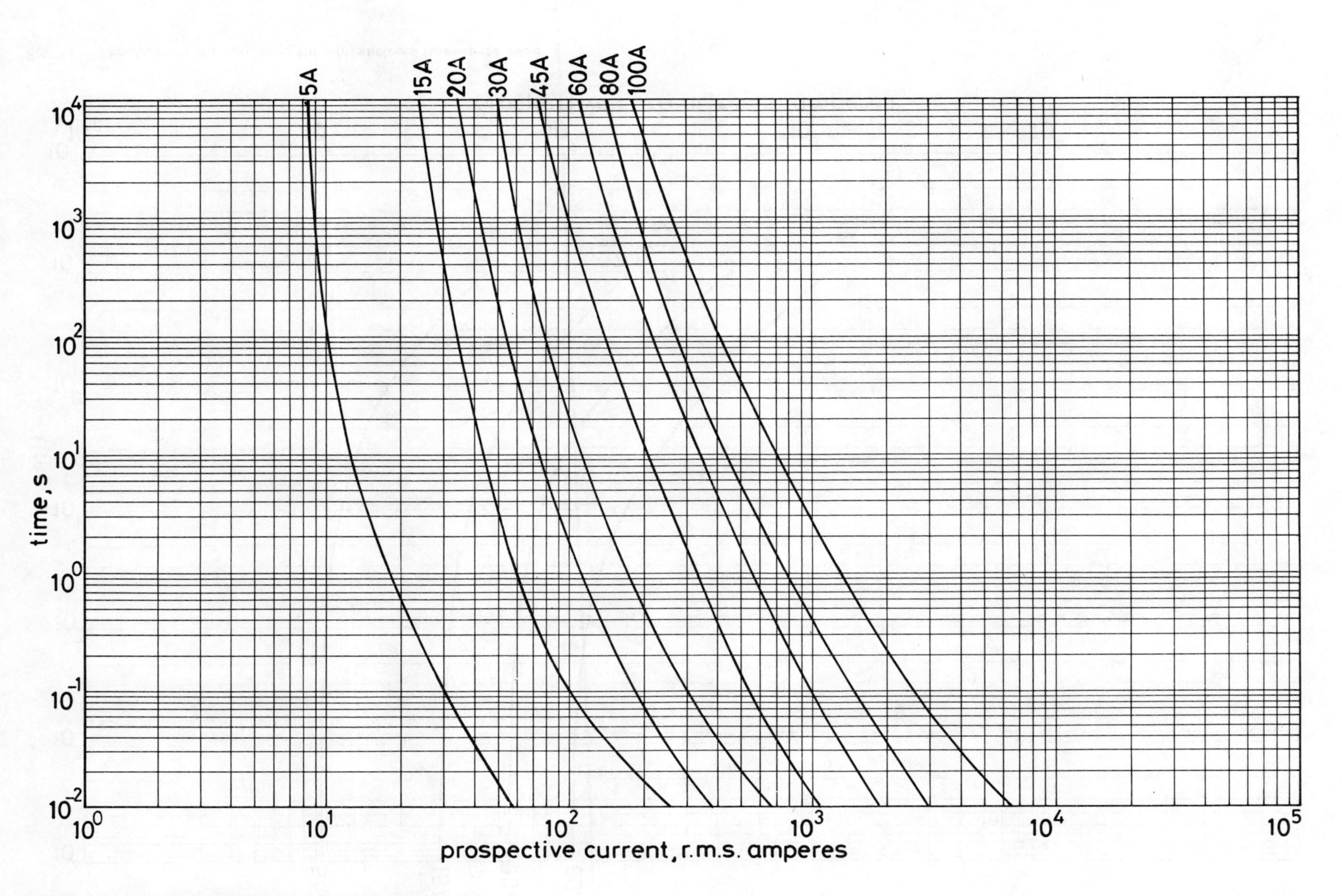

Fig. 10 Time/current characteristics for fuses to BS 1361

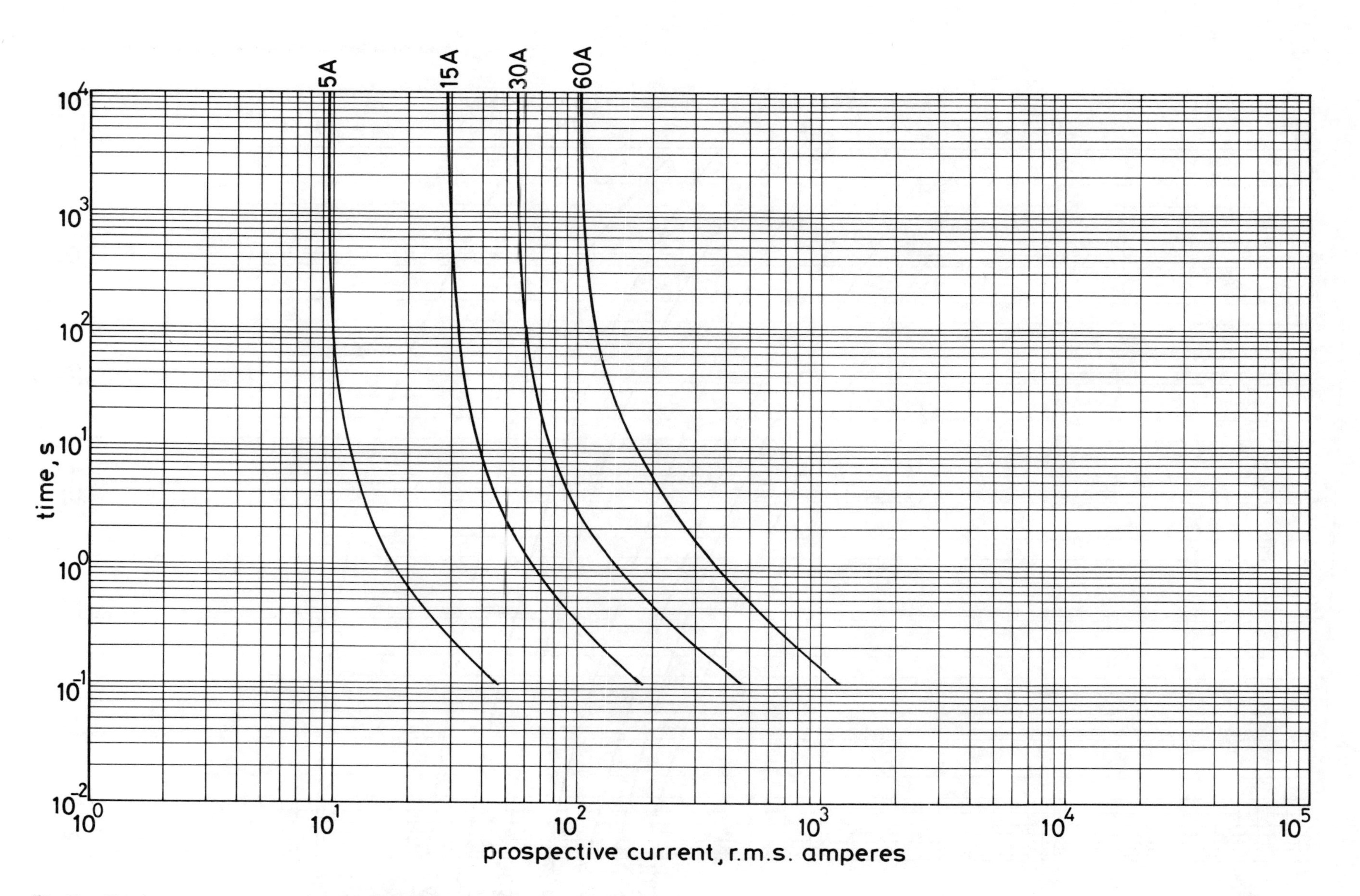

Fig. 11 Time/current characteristics for semi-enclosed fuses to BS 3036

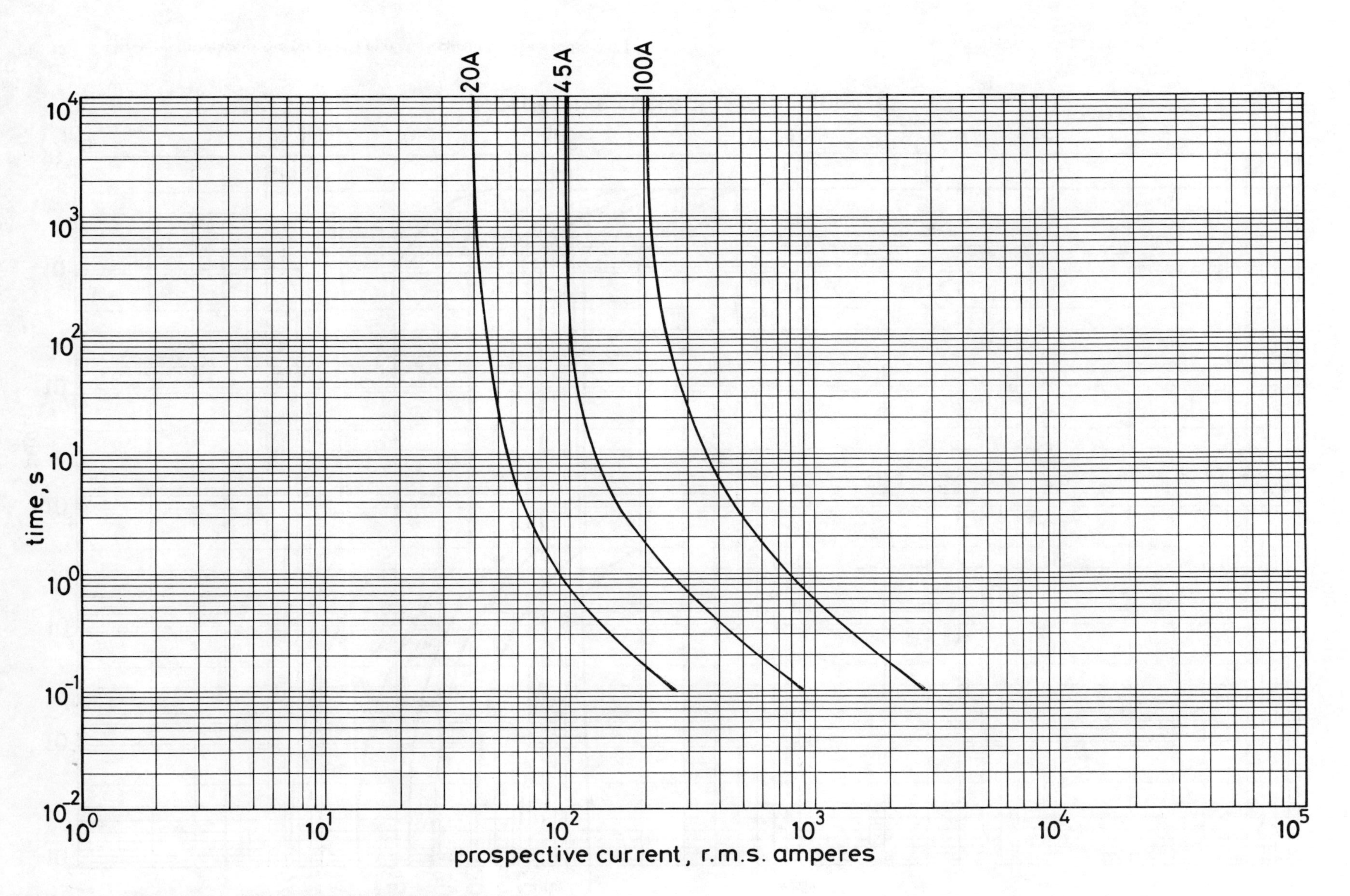

Fig. 12 Time/current characteristics for semi-enclosed fuses to BS 3036

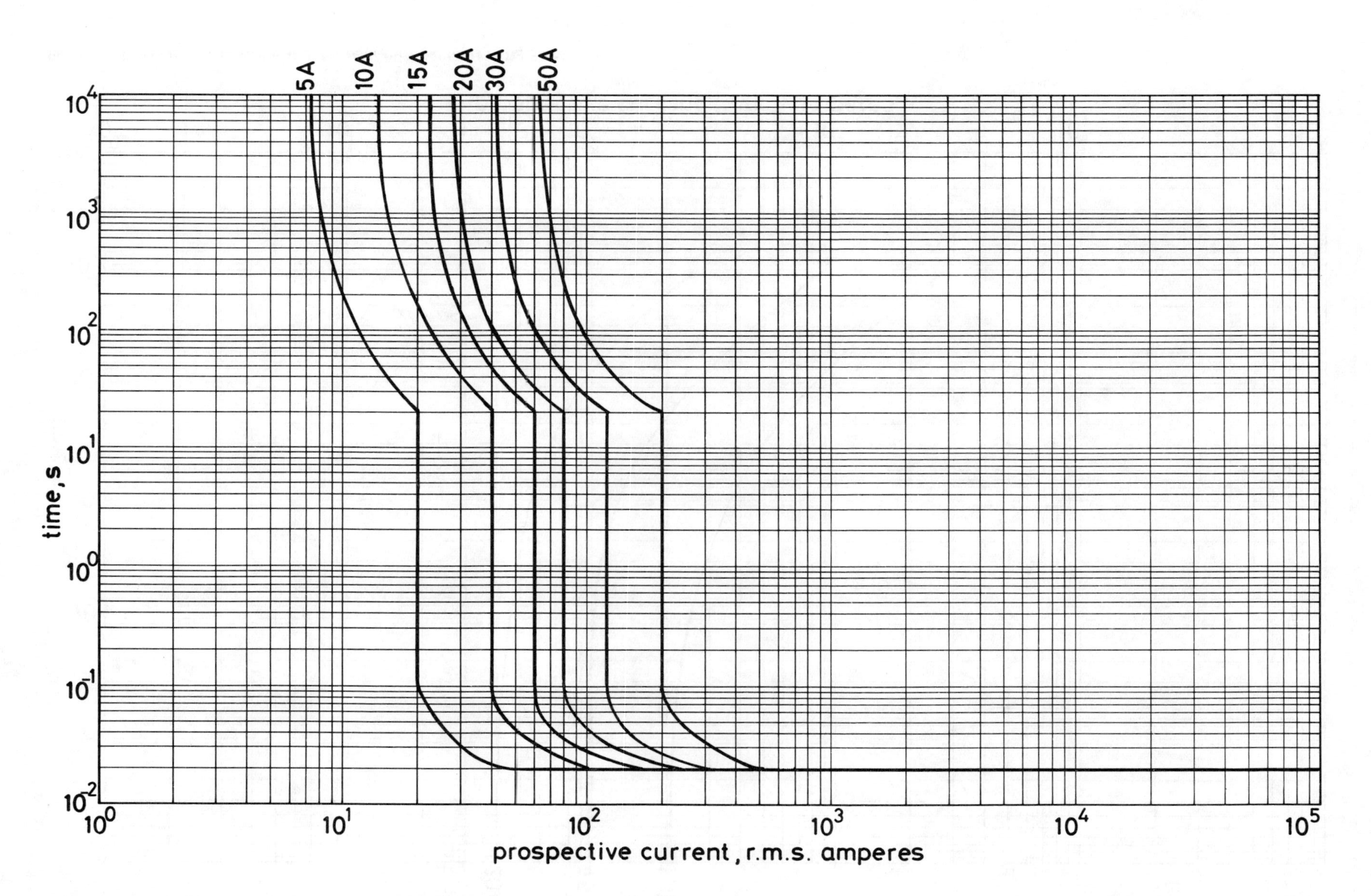

Fig. 13 Time/current characteristics for type 1 miniature circuit breakers to BS 3871

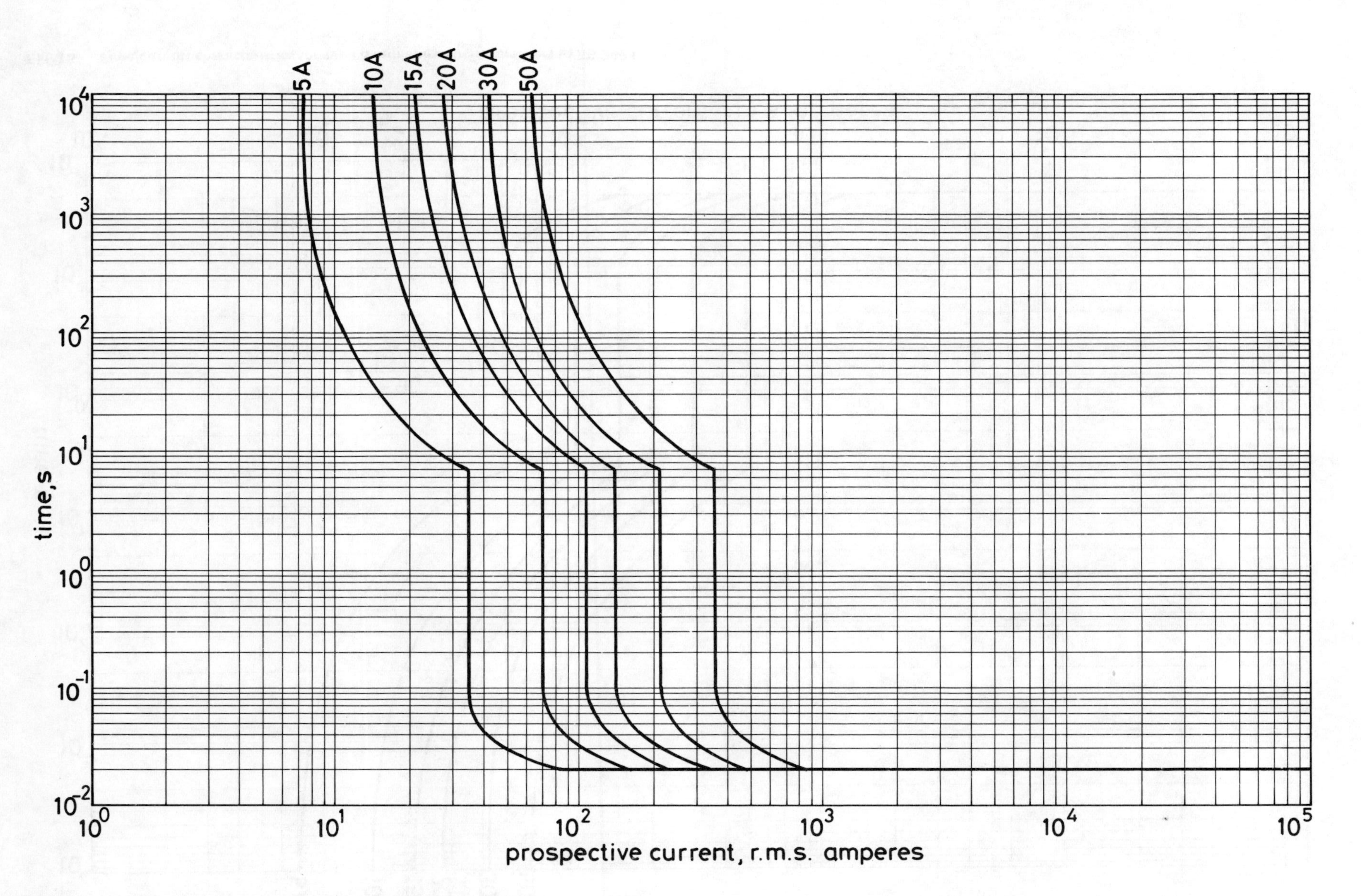

Fig. 14 Time/current characteristics for type 2 miniature circuit breakers to BS 3871

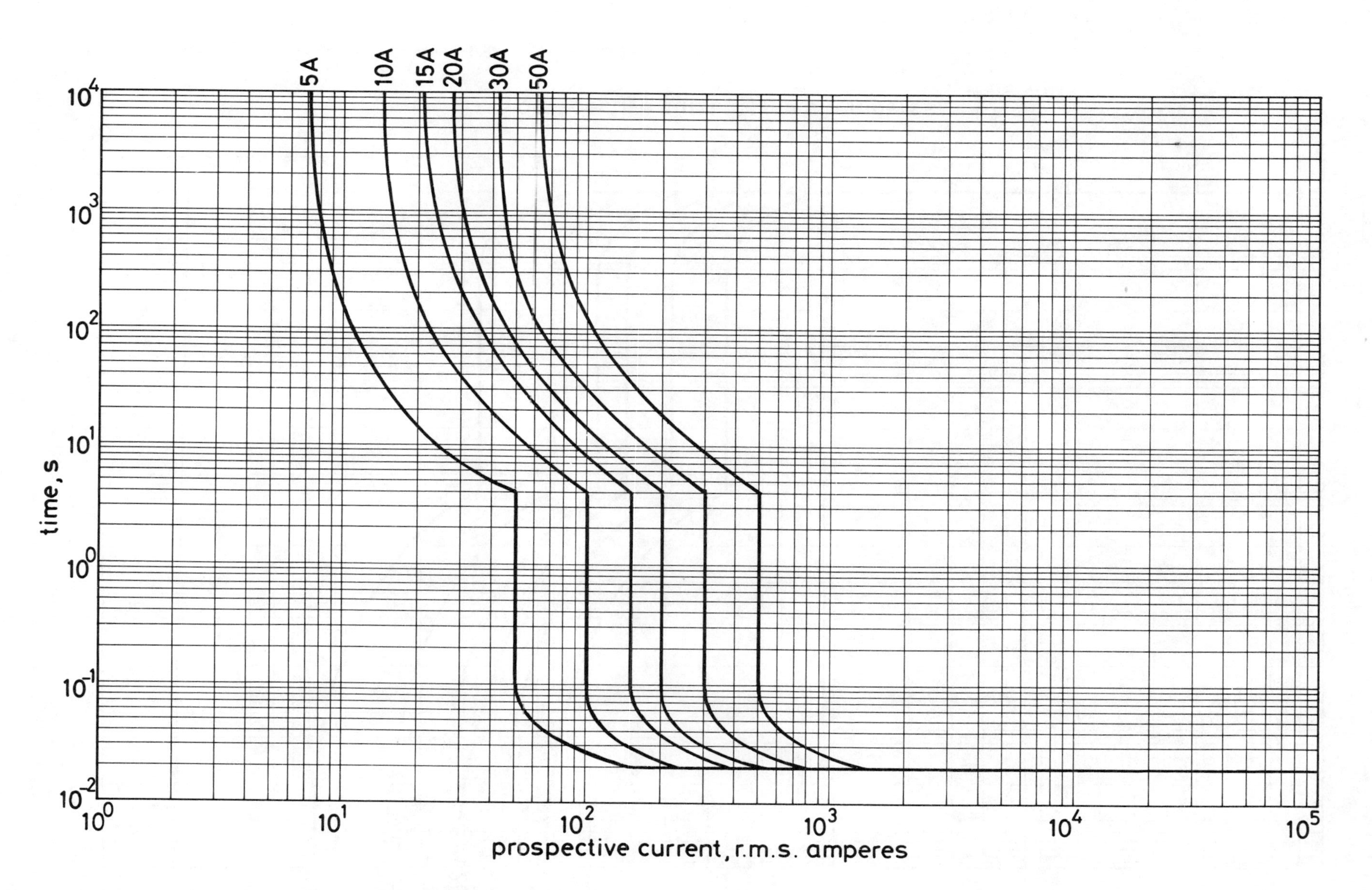

Fig. 15 Time/current characteristics for type 3 miniature circuit breakers to BS 3871

TABLE 8A

MAXIMUM EARTH FAULT LOOP IMPEDANCE (in ohms) WHEN OVERCURRENT PROTECTIVE DEVICE IS A FUSE TO BS 3036

(i) For circuits feeding socket outlets

Protective conductor mm^2	Fuse rating, amperes				
	5	15	20	30	45
1	9.6	2.7	1.8	1.2 (-)	–
1.5	9.6	2.7	1.8	1.1	–
2.5	9.6	2.7	1.8	1.1	0.6
4	9.6	2.7	1.8	1.1	0.6
6	9.6	2.7	1.8	1.1	0.6
10	9.6	2.7	1.8	1.1	0.6
16	9.6	2.7	1.8	1.1	0.6

(ii) For circuits feeding fixed equipment

Protective conductor mm^2	Fuse rating, amperes				
	5	15	20	30	45
1	20	5.6	4.0 (3.4)	1.2 (-)	–
1.5	20	5.6	4.0	2.8 (2.5)	–
2.5	20	5.6	4.0	2.8	1.6 (1.4)
4	20	5.6	4.0	2.8	1.6
6	20	5.6	4.0	2.8	1.6
10	20	5.6	4.0	2.8	1.6
16	20	5.6	4.0	2.8	1.6

The values given in brackets apply when the protective conductor is incorporated in the same p.v.c.-insulated cable as the associated phase conductors(s).

TABLE 8B

MAXIMUM EARTH FAULT LOOP IMPEDANCE (in ohms) WHEN OVERCURRENT PROTECTIVE DEVICE IS A FUSE TO BS 88 Pt. 2

(i) For circuits feeding socket outlets

Protective Conductor mm^2	Fuse rating, amperes							
	6	10	16	20	25	32	40	50
1	8.7	5.3	2.8	1.8	1.5	1.1 (0.92)	0.63 (0.5)	0.34 (0.28)
1.5	8.7	5.3	2.8	1.8	1.5	1.1	0.8 (0.77)	0.54 (0.43)
2.5	8.7	5.3	2.8	1.8	1.5	1.1	0.8	0.6
4	8.7	5.3	2.8	1.8	1.5	1.1	0.8	0.6
6	8.7	5.3	2.8	1.8	1.5	1.1	0.8	0.6
10	8.7	5.3	2.8	1.8	1.5	1.1	0.8	0.6
16	8.7	5.3	2.8	1.8	1.5	1.1	0.8	0.6

(ii) For circuits feeding fixed equipment

Protective conductor mm^2	Fuse rating, amperes							
	6	10	16	20	25	32	40	50
1	13.0	7.7	4.4	2.5 (2.2)	1.7 (1.5)	1.1 (0.9)	0.63 (0.6)	0.34 (0.28)
1.5	13.0	7.7	4.4	3.0 (2.9)	2.2 (2.0)	1.6 (1.4)	0.95 (0.8)	0.55 (0.44)
2.5	13.0	7.7	4.4	3.0	2.4	1.8	1.3 (1.1)	0.86 (0.75)
4	13.0	7.7	4.4	3.0	2.4	1.8	1.4	1.1
6	13.0	7.7	4.4	3.0	2.4	1.8	1.4	1.1
10	13.0	7.7	4.4	3.0	2.4	1.8	1.4	1.1
16	13.0	7.7	4.4	3.0	2.4	1.8	1.4	1.1

The values given in brackets apply where the protective conductor is incorporated in the same p.v.c.-insulated cable as the associated phase conductor(s).

TABLE 8C

MAXIMUM EARTH FAULT LOOP IMPEDANCE (IN OHMS) WHEN OVERCURRENT PROTECTIVE DEVICE IS A FUSE TO BS 1361

(i) For circuits feeding socket outlets

Protective conductor mm^2	Fuse rating, amperes				
	5	15	20	30	45
1	11.4	3.4	1.8	1.2 (1.1)	0.34 (0.28)
1.5	11.4	3.4	1.8	1.2	0.53 (0.44)
2.5	11.4	3.4	1.8	1.2	0.6
4	11.4	3.4	1.8	1.2	0.6
6	11.4	3.4	1.8	1.2	0.6
10	11.4	3.4	1.8	1.2	0.6
16	11.4	3.4	1.8	1.2	0.6

(ii) For circuits feeding fixed equipment

Protective conductor mm^2	Fuse rating, amperes				
	5	15	20	30	45
1	17	5.3	2.5 (2.4)	1.4 (1.1)	0.34 (0.28)
1.5	17	5.3	2.9	1.7 (1.5)	0.53 (0.44)
2.5	17	5.3	2.9	2.0	0.75 (0.65)
4	17	5.3	2.9	2.0	1.0 (0.90)
6	17	5.3	2.9	2.0	1.0
10	17	5.3	2.9	2.0	1.0
16	17	5.3	2.9	2.0	1.0

The values given in brackets apply when the protective conductor is incorporated in the same p.v.c.-insulated cable as the associated phase conductor(s).

TABLE 8D

Values of $(R_1 + R_2)$ per metre for p.v.c.-insulated copper conductors

Cross-sectional area, mm^2		$(R_1 + R_2)$ ohms/metre
Phase conductor	Protective conductor	
1	1	0.055
1.5	1	0.046
	1.5	0.037
2.5	1	0.039
	1.5	0.030
	2.5	0.022
4	1.5	0.026
	2.5	0.018
	4	0.014
6	2.5	0.016
	4	0.0116
	6	0.0092
10	4	0.0098
	6	0.0074
	10	0.0055
16	6	0.0064
	10	0.0045
	16	0.0035

APPENDIX 9

CURRENT-CARRYING CAPACITIES AND VOLTAGE DROPS FOR CABLES AND FLEXIBLE CORDS

CONTENTS

Preface to the tables.

Tables:

9A Typical methods of installation of cables and conductors.

9B Correction factors for groups of more than three single-core cables or more than one multicore cable.

9C Correction factors for cables in enclosed trenches.

COPPER CONDUCTORS

PVC-INSULATED CABLES

9D1 Single-core non-armoured, with or without sheath.

9D2 Twin and multicore non-armoured.

9D3 Twin and multicore armoured.

CABLES HAVING THERMOSETTING INSULATION

9E1 Twin and multicore armoured.

85°C RUBBER-INSULATED CABLES

9F1 Single-core.

PAPER-INSULATED CABLES

9G1 Single-core.

9G2 Twin and multicore.

FLEXIBLE CABLES AND CORDS

9H1 60°C rubber-insulated flexible cables.

9H2 85°C and 150°C rubber-insulated flexible cables.

9H3 Flexible cords.

MINERAL-INSULATED CABLES

9J1 Light duty, copper conductors and sheath, exposed to touch or p.v.c.-covered.

9J2 Light duty, copper conductors and sheath, having sheath bare and not exposed to touch.

9J3 Heavy duty, copper conductors and sheath, exposed to touch or p.v.c.-covered.

9J4 Heavy duty, copper conductors and sheath, having sheath bare and not exposed to touch.

9J5 Copper conductors and sheath, as earthed concentric wiring.

9J6 Light duty, copper conductors and aluminium sheath.

ALUMINIUM CONDUCTORS

PVC-INSULATED CABLES

9K1 Single-core, non-armoured.

9K2 Twin and multicore, non-armoured.

9K3 Twin and multicore, armoured.

CABLES HAVING THERMOSETTING INSULATION

9L1 Twin and multicore armoured.

PAPER-INSULATED CABLES

9M1 Single-core.

9M2 Twin and multicore.

MINERAL-INSULATED CABLES

9N1 Heavy duty, aluminium conductors and aluminium sheath.

APPENDIX 9

CURRENT-CARRYING CAPACITIES AND VOLTAGE DROPS FOR CABLES AND FLEXIBLE CORDS

1. Basis of tables

Current carrying capacities

At the time of issue of these Regulations, current-carrying capacities of conductors for IEC Publication 364 – 'Electrical installations of buildings' are under consideration. This appendix will be revised as soon as possible to take account of the results of the international work.

For the time being, therefore, the current-carrying capacities set out in this appendix are based on data provided by ERA Technology Ltd. and the Electric Cable Makers' Confederation (see also ERA Report 69-30 – 'Current rating standards for distribution cables'*).

The tabulated current-carrying capacities correspond to continuous loading and are also known as the 'full thermal current ratings' of the cables, corresponding to the conductor operating temperatures indicated in the headings to the tables. Cables may be seriously damaged, leading to early failure, or their service lives may be significantly reduced, if they are operated for any prolonged periods at temperatures above those corresponding to the tabulated current-carrying capacities. This is a major reason for the requirements of Regulation 433–2 concerning the relationship between the current-carrying capacities of conductors and the operating characteristics of overload protective devices.

The tabulated current-carrying capacities are based upon an ambient air temperature of 30°C. For other values of ambient air temperature it is necessary to apply a correction factor (multiplier) to obtain the corresponding effective current-carrying capacity.

In practice the ambient air temperature may be determined by thermometers placed in free air as close as practicable to the position at which the cables are installed or are to be installed, subject to the proviso that the measurements are not to be influenced by the heat arising from the cables; thus if the measurements are made while the cables are loaded, the thermometers must be placed about 0.5m or ten times the overall diameter of the cable, whichever is the lesser, from the cables, in the horizontal plane, or 150mm below the lowest of the cables.

The current-carrying capacities given in the tables for a.c. operation apply only to a frequency of 50 Hz. In extreme cases, notably for large multicore cables, the reduction in rating of cables carrying balanced 400 Hz a.c., compared with the rating at 50 Hz, may be as much as 25%. The effect on voltage drop may be considerably more than this for the larger sizes of cable, particularly where single-core cables in conduit are used. For small cables and flexible cords, such as may be used to supply individual tools, the difference in the 50 Hz and the 400 Hz current-carrying capacities may be negligible.

Voltage drop

Values of voltage drop are tabulated for a current of one ampere for a metre run, i.e. for a distance of 1m along the route taken by the cables, and represent the result of the voltage drops in all the circuit conductors. For any given run the values need to be multiplied by the length of the run in metres and by the current the cables are to carry, in amperes. For three-phase circuits the values tabulated relate to the line voltage and balanced load conditions have been assumed. Where the actual current to be carried, i.e. the design value of current for the circuit, is significantly less than the tabulated current-carrying capacity, the total voltage drop obtained by the above method is only approximate because the conductor temperature and hence its resistance will be less than those applicable at the tabulated current. Thus for a more accurate assessment, due allowance should be made for the change in conductor resistance with operating temperature.

The situation outlined in the previous paragraph will, for instance, occur when the conductor size is determined to allow for the use of a semi-enclosed fuse as the overcurrent protective device (see item 4(ii) below).

The tabulated values of voltage drop relate to the worst conditions, namely, where the phase angle of the cable circuit is equal to that of the load. For cables up to and including 120 mm^2 they apply with sufficient accuracy where the power factor of the load lies between 0.6 lagging and unity and, for larger cables, where the power factor of the load is not worse than 0.8 lagging. In all other cases the value may be unduly conservative and more exact calculation is necessary.

*Obtainable from ERA Technology Ltd., Cleeve Road, Leatherhead, Surrey KT22 7SA.

The tabulated values of voltage drop relate to the numbers and sizes of wires of conductors most commonly used in cables for fixed wiring of installations.

2. Types of cable and conditions of installation not provided for in the Tables

To reduce the number of tables that would otherwise need to be included, current-carrying capacities are not tabulated for types of cable and/or conditions of installation in less common use.

Information on some of these cables and conditions of installation will be provided by the Secretary of the Institution on receipt of an application stating details of the particular cable and installation concerned.

3. Determination of current-carrying capacity

In order to determine the current-carrying capacity of a cable in a particular installation it may be necessary to apply one or more correction factors to the tabulated value given in the appropriate table for that cable.

(i) *For ambient temperature*

Each table gives the correction factor to be applied dependent on the actual ambient temperature in the installation.

NOTE – IF FUSES TO BS 3036 ARE USED THE AMBIENT TEMPERATURE CORRECTION FACTORS TO BE APPLIED ARE GIVEN IN ITEM 4 BELOW.

(ii) *For grouping*

Where a correction factor for grouping has to be applied, see Table 9B.

(iii) *For thermal insulation*

See Note to Regulation 522–6.

4. Determination of the size of cable to be used

Having established the design current of the circuit under consideration, and having chosen the type and nominal current or current setting in accordance with Regulation 433–2 of the overcurrent protective device it is intended to use, the following procedure enables the designer to determine the size of cable it will be necessary to use in order to comply with the requirements for overload protection.

(i) *Where the protective device is a fuse to BS 88 or BS 1361 or a circuit breaker to BS 3871 Part 2 or BS 4572 Part 2*

- DIVIDE the nominal current of the protective device by the appropriate ambient temperature correction factor given in the table for the type of cable it is intended to use.
- Then further DIVIDE by any applicable correction factor for grouping given in Table 9B.
- The size of cable to be used is such that its tabulated current-carrying capacity for the installation method concerned is not less than the value of nominal current of the protective device adjusted as above.

(ii) *Where the protective device is a semi-enclosed fuse to BS 3036*

- DIVIDE the nominal current of the protective device by the appropriate ambient temperature correction factor given in the following table.

Ambient temperature °C	25	30	35	40	45	50	55	60	65	70	75	80
PVC-insulated	1.02	1.0	0.97	0.94	0.91	0.88	0.77	0.63	0.44	–	–	–
85°C rubber-insulated	1.02	1.0	0.97	0.94	0.92	0.89	0.85	0.77	0.68	0.59	0.47	0.33

- Then further DIVIDE by any applicable factor for grouping given in Table 9B.
- Then further DIVIDE by 0.725.
- The size of cable to be used is such that its tabulated current-carrying capacity is not less than the value of nominal current of the protective device adjusted as above.

5. Methods of installation of cables and conductors

Table 9A lists methods of installation in common use but, as stated in Regulation 521–13, the use of other methods is not precluded where specified by a suitably qualified electrical engineer. In the latter case the evaluation of current-carrying capacity may need to be based on experimental work.

TABLE 9A

Typical methods of installation of cables and conductors

'Enclosed'

Type	Description	Examples
A	Single-core and multicore cables, enclosed in conduit.	
B	Single-core and multicore cables enclosed in cable trunking.	
C	Single-core and multicore cables enclosed in underground conduit, ducts, and cable ducting.	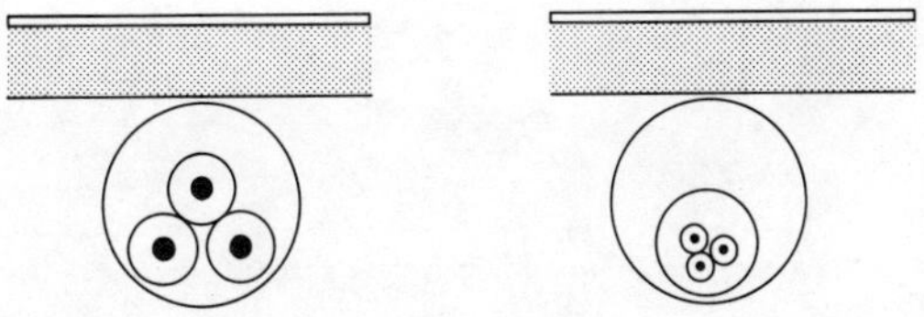
D	Two or more single-core cables contained in separate bores of a multicore conduit and intended to be solidly embedded in concrete or plaster or generally incorporated in the building structure (may be used as a prefabricated wiring system).	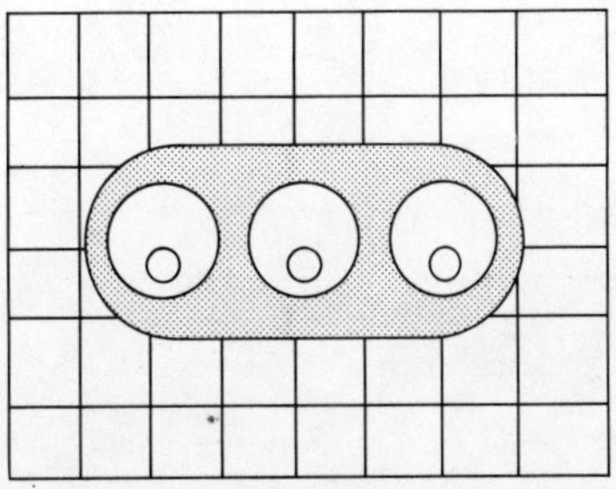

TABLE 9A (contd.)

'Open and clipped direct'

E	Sheathed single-core and multicore cables clipped direct to or lying on a non-metallic surface.	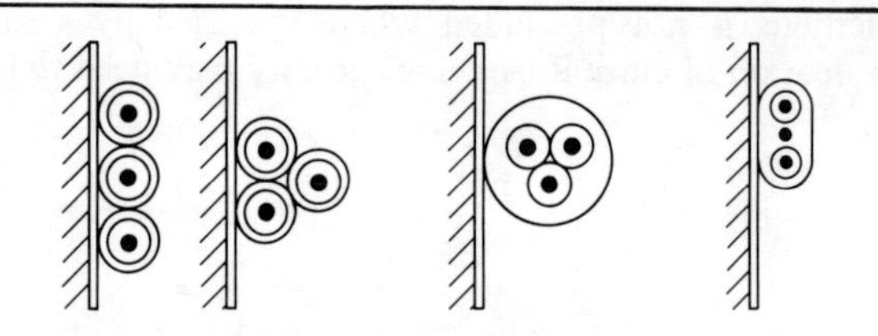 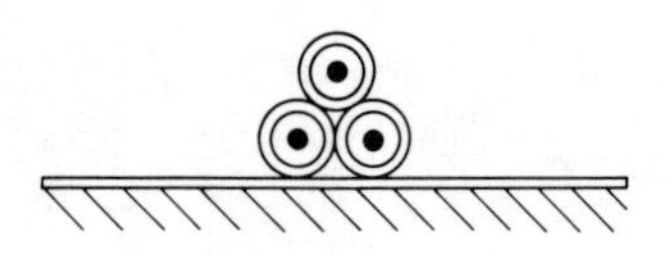
F	Sheathed single-core and multicore cables on a cable tray, bunched and unenclosed.	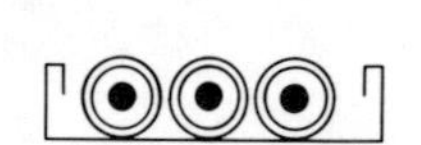 
G	Sheathed cables embedded direct in plaster other than special thermally insulating plasters.	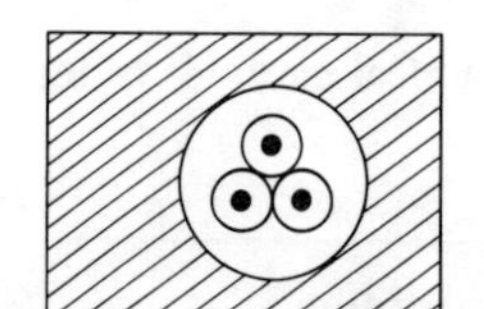 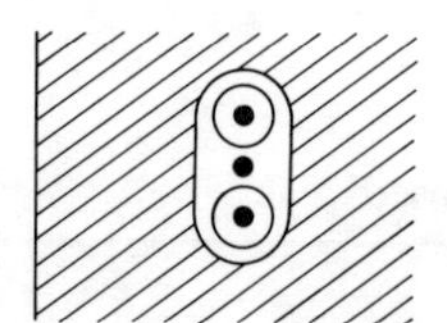
H	Sheathed single-core and multicore cables suspended from or incorporating a catenary wire.	 

TABLE 9A (contd.)

'Defined conditions'

J Sheathed single-core cables in free air.

Vertical surface of a wall or open cable trench

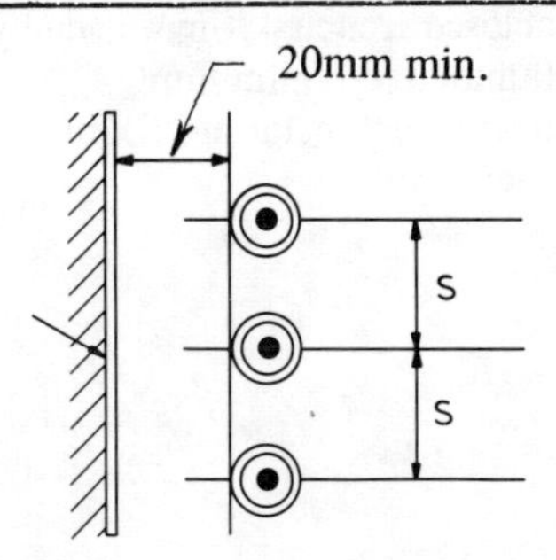

For cables in which the conductor cross-sectional area does not exceed 185 mm² S is equal to twice the overall diameter of the cable. For cables in which the conductor cross-sectional area exceeds 185 mm², S is about 90mm. For two cables in horizontal formation on brackets fixed to a wall, S may have any lesser value.

Vertical surface of a wall or open cable trench

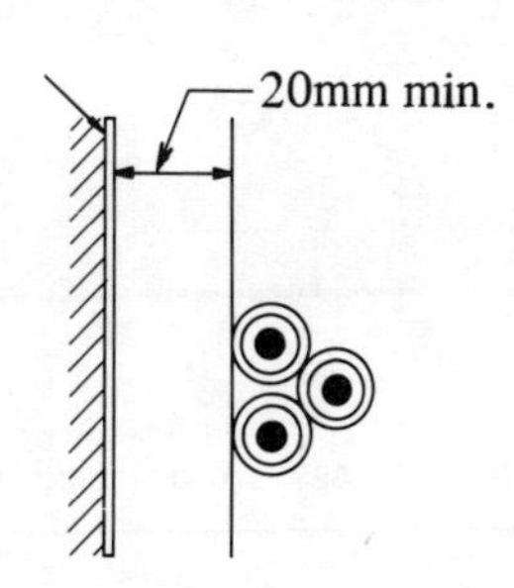

The values given apply provided that the sheaths of single-core metal-sheathed cables are electrically bonded at each end of the run. The cables are assumed to be remote from iron, steel or ferro-concrete other than the cable supports.

Cables spaced by distances less than those described above are assumed to be 'clipped direct' (see Method E).

K Sheathed twin and multicore cables in free air.

Vertical surface of a wall or open cable trench.

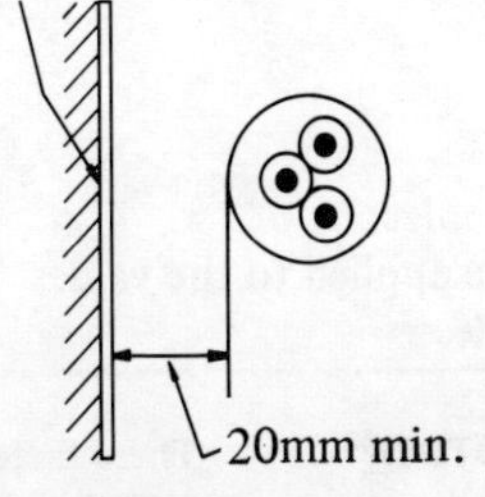

Cables spaced by a lesser distance are assumed to be 'clipped direct' (see Method E).

TABLE 9A (contd.)

In enclosed trenches

L	Single and multicore cables in enclosed trench 450mm wide by 300mm deep (minimum dimensions) including 100mm cover.	Two single-core cables with surfaces separated by a distance equal to one diameter; three single-core cables in trefoil and touching throughout. Multicore cables or groups of single-core cables separated by a minimum distance of 50mm.	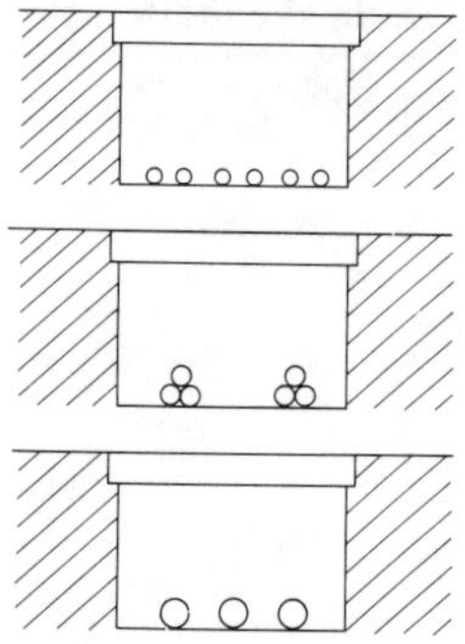
M	Single and multicore cables in enclosed trench 450mm wide by 600mm deep (minimum dimensions) including 100mm cover.	Single-core cables arranged in flat groups of two or three on the vertical trench wall with surfaces separated by a distance equal to one diameter with a minimum* separation of 50mm between groups. Multicore cables installed singly separated by a minimum* distance of 75mm. All cables spaced at least 25mm from the trench wall.	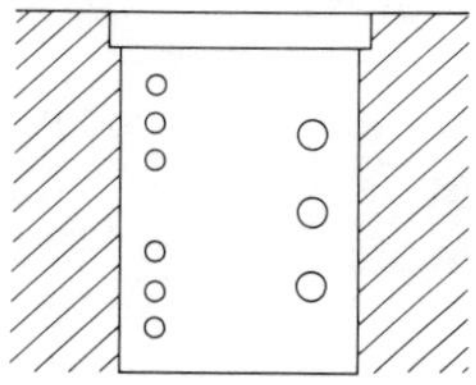
N	Single and multicore cables in enclosed trench 600mm wide by 760mm deep (minimum dimensions) including 100mm cover.	Single-core cables arranged in groups of two or three in flat formation with the surfaces separated by a distance equal to one diameter or in trefoil formation with cables touching Groups separated by a minimum* distance of 50mm either horizontally or vertically. Multicore cables installed singly separated by a minimum* distance of 75mm either horizontally or vertically. All cables spaced at least 25mm from the trench wall.	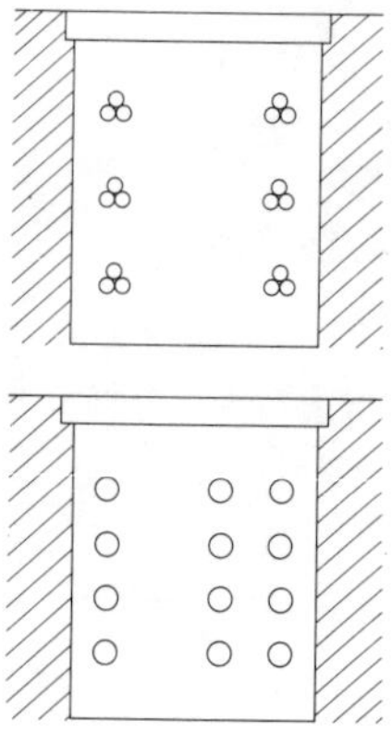

*Larger spacings to be used where practicable.

TABLE 9B

Correction factors for groups of more than three single-core cables or more than one multicore cable

Type of cable and installation condition	Number of loaded conductors											
	4	6	8	10	12	16	20	24	28	32	36	40
Single-core cables: Factor to be applied to the values for two single-core cables	0.80	0.69	0.62	0.59	0.55	0.51	0.48	0.43	0.41	0.39	0.38	0.36
	Number of cables											
	2	3	4	5	6	8	10	12	14	16	18	20
Multicore cables: Factor to be applied to the values for one cable	0.80	0.70	0.65	0.60	0.57	0.52	0.48	0.45	0.43	0.41	0.39	0.38

NOTES:

1. These factors are applicable to groups of cables all of one size, equally loaded, including groups bunched in more than one plane.
2. Where spacing between adjacent cables exceeds twice their overall diameter, no reduction factor need be applied.

TABLE 9C

Correction factors for cables installed in enclosed trenches (installation methods L, M and N of Table 9A)

The correction factors tabulated below relate to the dispositions of cables illustrated in items L, M and N of Table 9A and are applicable to the current-carrying capacities and volt drops for installation methods J and K of Table 9A as given in the following tables of this appendix.

Nominal cross-sectional area of conductor	Correction factors									
	Type L of Table 9A				Type M of Table 9A			Type N of Table 9A		
	Two single-core cables, or one 3- or 4-core cable	Three single-core cables, or two twin cables	Four single-core cables, or two 3- or 4-core cables	Six single-core cables, four twin cables, or three 3- or 4-core cables	Six single-core cables, four twin cables, or three 3- or 4-core cables	Eight single-core cables, or four 3- or 4-core cables	Twelve single-core cables, eight twin cables, or six 3- or 4-core cables	Twelve single-core cables, eight twin cables, or six 3- or 4-core cables	Eighteen single-core cables, twelve twin cables, or nine 3- or 4-core cables	Twenty-four single core cables sixteen twin cables, or twelve 3- or 4-core cables
1	2	3	4	5	6	7	8	9	10	11
mm^2										
4	0.93	0.90	0.87	0.82	0.86	0.83	0.76	0.81	0.74	0.69
6	0.92	0.89	0.86	0.81	0.86	0.82	0.75	0.80	0.73	0.68
10	0.91	0.88	0.85	0.80	0.85	0.80	0.74	0.78	0.72	0.66
16	0.91	0.87	0.84	0.78	0.83	0.78	0.71	0.76	0.70	0.64
25	0.90	0.86	0.82	0.76	0.81	0.76	0.69	0.74	0.67	0.62
35	0.89	0.85	0.81	0.75	0.80	0.74	0.68	0.72	0.66	0.60
50	0.88	0.84	0.79	0.74	0.78	0.73	0.66	0.71	0.64	0.59
70	0.87	0.82	0.78	0.72	0.77	0.72	0.64	0.70	0.62	0.57
95	0.86	0.81	0.76	0.70	0.75	0.70	0.63	0.68	0.60	0.55
120	0.85	0.80	0.75	0.69	0.73	0.68	0.61	0.66	0.58	0.53
150	0.84	0.78	0.74	0.67	0.72	0.67	0.59	0.64	0.57	0.51
185	0.83	0.77	0.73	0.65	0.70	0.65	0.58	0.63	0.55	0.49
240	0.82	0.76	0.71	0.63	0.69	0.63	0.56	0.61	0.53	0.48
300	0.81	0.74	0.69	0.62	0.68	0.62	0.54	0.59	0.52	0.46
400	0.80	0.73	0.67	0.59	0.66	0.60	0.52	0.57	0.50	0.44
500	0.78	0.72	0.66	0.58	0.64	0.58	0.51	0.56	0.48	0.43
630	0.77	0.71	0.65	0.56	0.63	0.57	0.49	0.54	0.47	0.41

COPPER CONDUCTORS

TABLE 9D1

Current-carrying capacities and associated voltage drops for single-core p.v.c.-insulated cables, non-armoured, with or without sheath (copper conductors)

BS 6004
BS 6346

Conductor operating temperature: 70°C

Conductor cross-sectional area	Installation methods A to C† of Table 9A ('Enclosed')				Installation methods E to H of Table 9A ('Clipped direct')				Installation method J of Table 9A ('Defined conditions')					
	2 cables, single-phase a.c., or d.c.		3 or 4 cables, three-phase a.c.		2 cables, single-phase a.c., or d.c.		3 or 4 cables, three-phase a.c.		Flat or vertical (2 cables, single-phase a.c., or d.c., or 3 or 4 cables three-phase)				Trefoil (3 cables three-phase)	
	Current carrying capacity	Volt drop per ampere per metre	Current carrying capacity	Volt drop per ampere per metre	Current carrying capacity	Volt drop per ampere per metre	Current carrying capacity	Volt drop per ampere per metre	Current carrying capacity	Volt drop per ampere per metre: Single-phase	Volt drop per ampere per metre: d.c.	Volt drop per ampere per metre: Three-phase	Current carrying capacity	Volt drop per ampere per metre
1	2	3	4	5	6	7	8	9	10	11	12	13	14	15
mm²	A	mV	A	mV	A	mV	A	mV	A	mV	mV	mV	A	mV
1.0	14	42	12	37	17	42	16	37	–	–	–	–	–	–
1.5	17	28	14	24	21	28	20	24	–	–	–	–	–	–
2.5	24	17	21	15	30	17	26	15	–	–	–	–	–	–
4	32	11	29	9.2	40	11	36	9.2	–	–	–	–	–	–
6	41	7.1	37	6.2	50	7.1	45	6.2	–	–	–	–	–	–
10	55	4.2	51	3.7	68	4.2	61	3.7	–	–	–	–	–	–
16	74	2.7	66	2.3	90	2.7	81	2.3	–	–	–	–	–	–
25	97	1.7	87	1.5	118	1.7	106	1.5	–	–	–	–	–	–
35	119	1.3	106	1.1	145	1.3	130	1.1	–	–	–	–	–	–
		a.c. d.c.				a.c. d.c.								
50	145	0.97 0.91	125	0.84	175	0.93 0.91	160	0.82	195	0.95	0.91	0.85	170	0.80
70	185	0.71 0.63	160	0.62	220	0.65 0.63	200	0.59	240	0.68	0.63	0.62	210	0.59
95	230	0.56 0.45	195	0.48	270	0.48 0.45	240	0.45	300	0.52	0.45	0.49	260	0.42
120	260	0.48 0.36	220	0.42	310	0.40 0.36	280	0.38	350	0.44	0.36	0.43	300	0.34
150	–	– –	–	–	355	0.34 0.29	320	0.34	410	0.39	0.29	0.39	350	0.29
185	–	– –	–	–	405	0.29 0.24	365	0.30	470	0.35	0.24	0.36	400	0.25
240	–	– –	–	–	480	0.24 0.18	430	0.27	560	0.36	0.18	0.38	480	0.22
300	–	– –	–	–	560	0.22 0.14	500	0.25	660	0.33	0.14	0.35	570	0.19
400	–	– –	–	–	680	0.20 0.12	610	0.24	800	0.30	0.12	0.33	680	0.17
500	–	– –	–	–	800	0.18 0.086	710	0.23	910	0.28	0.086	0.31	770	0.16
630	–	– –	–	–	910	0.17 0.068	820	0.22	1040	0.26	0.068	0.30	880	0.15

†For installation method C, the tabulated values are applicable only to the range up to and including 35mm². For larger sizes in this installation method, see ERA Report 69–30. For cables in ducts in the floor of a building, the ERA ratings must be adjusted by the appropriate factor for ambient temperature. For installation method D, the current-carrying capacities for twin and multicore sheathed cables in methods A to C, up to 35mm², are applicable (see Table 9D2).

NOTES: 1 – **WHERE THE CONDUCTOR IS TO BE PROTECTED BY A SEMI-ENCLOSED FUSE TO BS 3036, SEE ITEM 4(ii) OF THE PREFACE TO THIS APPENDIX.**

2 – The current-carrying capacities in columns 6 and 8 are applicable to flexible cables to BS 6004 Table 1(b) when the cables are used in fixed installations.

CORRECTION FACTORS

FOR AMBIENT TEMPERATURE

Ambient temperature	25°C	35°C	40°C	45°C	50°C	55°C	60°C	65°C
Correction factor	1.06	0.94	0.87	0.79	0.71	0.61	0.50	0.35

FOR GROUPING

See Table 9B except that the factors for 32, 36 and 40 conductors do not apply to three-phase.

COPPER CONDUCTORS

TABLE 9D2

Current-carrying capacities and associated voltage drops for twin and multicore p.v.c.-insulated cables, non-armoured (copper conductors)

BS 6004
BS 6346

Conductor operating temperature: 70°C

Conductor cross-sectional area	Installation methods A to C† of Table 9A ('Enclosed')				Installation methods E to H of Table 9A ('Clipped direct')				Installation method K of Table 9A ('Defined conditions')			
	One twin cable, with or without protective conductor, single-phase a.c., or d.c.		One three-core cable, with or without protective conductor, or one four-core cable, three-phase		One twin cable, with or without protective conductor, single-phase a.c., or d.c.		One three-core cable, with or without protective conductor, or one four-core cable, three-phase		One twin cable, with or without protective conductor, single-phase a.c., or d.c.		One three-core cable, with or without protec-ive conductor or one four-core cable, three-phase	
	Current carrying capacity	Volt drop per ampere per metre	Current carrying capacity	Volt drop per ampere per metre	Current carrying capacity	Volt drop per ampere per metre	Current carrying capacity	Volt drop per ampere per metre	Current carrying capacity	Volt drop per ampere per metre	Current carrying capacity	Volt drop per ampere per metre
1	2	3	4	5	6	7	8	9	10	11	12	13
mm^2	A	mV	A	mV	A	mV	A	mV	A	mV	A	mV
1.0	14	42	12	37	16	42	13	37	–	–	–	–
1.5	18	28	16	24	20	28	17	24	–	–	–	–
2.5	24	17	21	15	28	17	24	15	–	–	–	–
4	32	11	29	9.2	36	11	32	9.2	–	–	–	–
6	40	7.1	36	6.2	46	7.1	40	6.2	–	–	–	–
10	53	4.2	49	3.7	64	4.2	54	3.7	–	–	–	–
16	70	2.7	62	2.3	85	2.7	71	2.3	–	–	–	–
25	79	1.8	70	1.6	108	1.8	90	1.6	114	1.8	95	1.6
35	98	1.3	86	1.1	132	1.3	115	1.1	139	1.3	122	1.1
50	–	–	–	–	163	0.92	140	0.81	172	0.92	148	0.81
						a.c. d.c.				a.c. d.c.		
70	–	–	–	–	207	0.65 0.64	176	0.57	218	0.65 0.64	186	0.57
95	–	–	–	–	251	0.48 0.46	215	0.42	265	0.48 0.46	227	0.42
120	–	–	–	–	290	0.40 0.36	251	0.34	306	0.40 0.36	265	0.34
150	–	–	–	–	330	0.32 0.25	287	0.29	348	0.32 0.25	302	0.29
185	–	–	–	–	380	0.29 0.23	330	0.24	400	0.29 0.23	348	0.24
240	–	–	–	–	450	0.25 0.18	392	0.20	474	0.25 0.18	413	0.20
300	–	–	–	–	520	0.23 0.14	450	0.18	548	0.23 0.14	474	0.18
400	–	–	–	–	600	0.22 0.11	520	0.17	632	0.22 0.11	548	0.17

FLAT CABLES ONLY (1.0 to 16 mm^2)

†For installation method C, the tabulated values are applicable only to the range up to and including 35 mm^2. For larger sizes in this installation method, see ERA Report 69–30. For cables in ducts in the floor of a building, the ERA ratings must be adjusted by the appopriate factor for ambient temperature.

NOTES: 1 – WHERE THE CONDUCTOR IS TO BE PROTECTED BY A SEMI–ENCLOSED FUSE TO BS 3036, SEE ITEM 4(ii) OF THE PREFACE TO THIS APPENDIX.

2 – The current carrying capacities in columns 6 and 8 are applicable to flexible cables to BS 6004 Table 1(b) where the cables are used in fixed installations.

CORRECTION FACTORS

FOR AMBIENT TEMPERATURE

Ambient temperature	25°C	35°C	40°C	45°C	50°C	55°C	60°C	65°C
Correction factor	1.06	0.94	0.87	0.79	0.71	0.61	0.50	0.35

FOR GROUPING
See Table 9B.

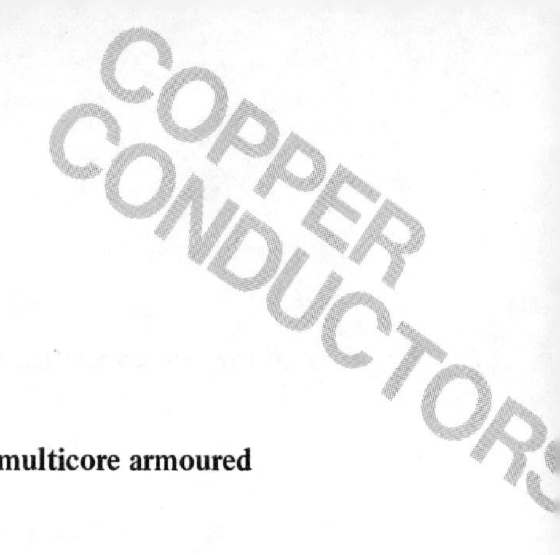

TABLE 9D3

Current-carrying capacities and associated voltage drops for twin and multicore armoured p.v.c.-insulated cables (copper conductors)

BS 6346

Conductor operating temperature: 70°C

Conductor cross-sectional area	Installation methods E, F and G† of Table 9A ('Clipped direct')				Installation method K of Table 9A ('Defined conditions')			
	One twin cable, single-phase a.c., or d.c.		One three- or four-core cable, three-phase		One twin cable, single-phase a.c., or d.c.		One three- or four-core cable, three-phase	
	Current carrying capacity	Volt drop per ampere per metre	Current carrying capacity	Volt drop per ampere per metre	Current carrying capacity	Volt drop per ampere per metre	Current carrying capacity	Volt drop per ampere per metre
1	2	3	4	5	6	7	8	9
mm²	A	mV	A	mV	A	mV	A	mV
1.5	20	29	18	25	–	–	–	–
2.5	29	18	24	16	–	–	–	–
4	37	12	31	9.6	–	–	–	–
6	48	7.4	41	6.3	50	7.3	42	6.3
10	66	4.3	56	3.8	69	4.3	58	3.8
16	86	2.7	73	2.3	90	2.7	77	2.3
25	115	1.8	97	1.6	121	1.8	102	1.6
35	142	1.3	119	1.1	149	1.3	125	1.1
50	168	0.92	147	0.81	180	0.92	155	0.81
		a.c. d.c.				a.c. d.c.		
70	209	0.65 0.64	180	0.57	220	0.65 0.64	190	0.57
95	257	0.48 0.46	219	0.42	270	0.48 0.46	230	0.42
120	295	0.40 0.36	257	0.34	310	0.40 0.36	270	0.34
150	337	0.32 0.25	295	0.29	355	0.32 0.25	310	0.29
185	390	0.29 0.23	333	0.24	410	0.29 0.23	350	0.24
240	461	0.25 0.18	399	0.20	485	0.25 0.18	420	0.20
300	523	0.23 0.14	451	0.18	550	0.23 0.14	475	0.18
400	589	0.22 0.11	523	0.17	620	0.22 0.11	550	0.17

†For installation method C, see ERA Report 69–30. For cables in ducts in the floor of a building, the ERA ratings must be adjusted by the appropriate factor for ambient temperature.

NOTE – WHERE THE CONDUCTOR IS TO BE PROTECTED BY A SEMI-ENCLOSED FUSE TO BS 3036, SEE ITEM 4 (ii) OF THE PREFACE TO THIS APPENDIX.

CORRECTION FACTORS

FOR AMBIENT TEMPERATURE

Ambient temperature	25°C	35°C	40°C	45°C	50°C	55°C	60°C	65°C
Correction factor	1.06	0.94	0.87	0.79	0.71	0.61	0.50	0.35

FOR GROUPING

See Table 9B.

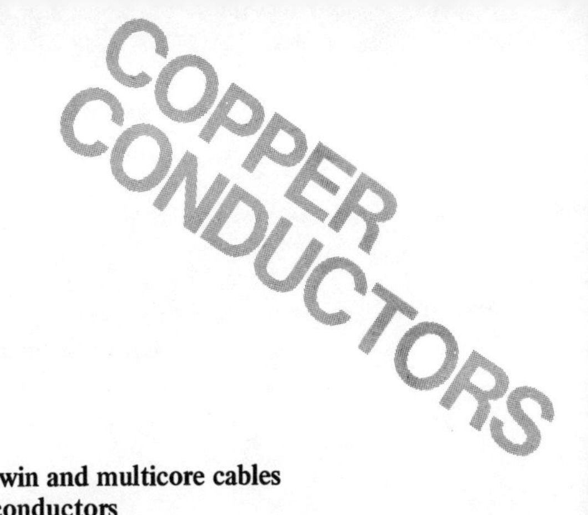

TABLE 9E1

Current-carrying capacities and associated voltage drops for armoured twin and multicore cables to BS 5467, having thermosetting insulation and copper conductors

Conductor operating temperature: 90°C

Conductor cross-sectional area	Installation methods E, F and G† of Table 9A ('Clipped direct')					Installation method K of Table 9A ('Defined conditions')				
	One twin cable, a.c. or d.c.			One three- or four-core cable, balanced three-phase a.c.		One twin cable, a.c. or d.c.			One three- or four-core cable, balanced three-phase a.c.	
	Current carrying capacity	Volt drop per ampere per metre a.c.	Volt drop per ampere per metre d.c.	Current carrying capacity	Volt drop per ampere per metre	Current carrying capacity	Volt drop per ampere per metre a.c.	Volt drop per ampere per metre d.c.	Current carrying capacity	Volt drop per ampere per metre
1	2	3	4	5	6	7	8	9	10	11
mm^2	A	mV	mV	A	mV	A	mV	mV	A	mV
16	108	2.9	2.9	95	2.6	114	2.9	2.9	100	2.6
25	144	1.9	1.9	126	1.6	152	1.9	1.9	133	1.6
35	181	1.3	1.3	153	1.2	190	1.3	1.3	160	1.2
50	217	1.0	0.99	185	0.87	228	1.0	0.99	195	0.87
70	271	0.70	0.68	235	0.61	285	0.70	0.68	247	0.61
95	338	0.52	0.49	289	0.45	356	0.52	0.49	304	0.45
120	388	0.42	0.39	334	0.36	408	0.42	0.39	350	0.36
150	442	0.35	0.32	388	0.30	465	0.35	0.32	408	0.30
185	514	0.29	0.25	442	0.25	540	0.29	0.25	465	0.25
240	605	0.24	0.19	523	0.21	635	0.24	0.19	550	0.21
300	695	0.21	0.15	596	0.19	730	0.21	0.15	627	0.19

†For installation method C, see ERA Report 69–30. For cables in ducts in the floor of a building, the ERA ratings must be adjusted by the appropriate factor for ambient temperature.

NOTE – WHERE THE CONDUCTOR IS TO BE PROTECTED BY A SEMI-ENCLOSED FUSE TO BS 3036, SEE ITEM 4(ii) OF THE PREFACE TO THIS APPENDIX.

CORRECTION FACTORS

FOR AMBIENT TEMPERATURE

Ambient Temperature	25°C	35°C	40°C	45°C	50°C	55°C	60°C	65°C	70°C	75°C	80°C
Correction factor	1.04	0.96	0.91	0.87	0.82	0.76	0.71	0.65	0.58	0.5	0.41

FOR GROUPING

See Table 9B.

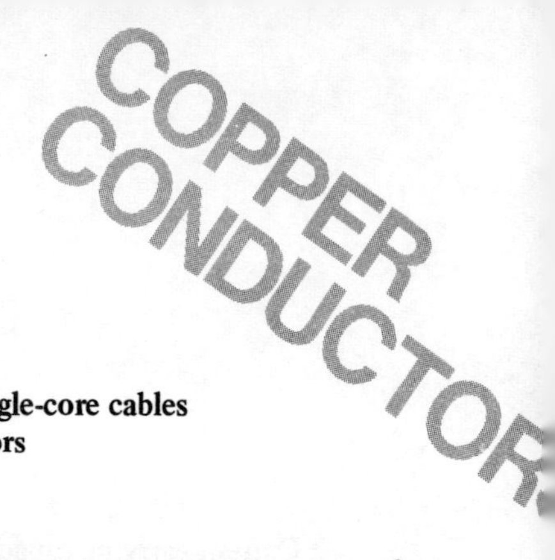

TABLE 9F1

Current-carrying capacities and associated voltage drops for single-core cables having 85°C rubber insulation and copper conductors

(BS 6007)

Conductor operating temperature: 85°C

Conductor cross-sectional area	Installation methods A to C† of Table 9A ('Enclosed')				Installation methods E to H of Table 9A ('Clipped direct')				Installation method J of Table 9A ('Defined conditions')			
	2 cables, single-phase a.c., or d.c.		3 or 4 cables, three-phase a.c.		2 cables, single-phase a.c., or d.c.		3 or 4 cables, three-phase a.c.		Flat or vertical (2 cables, single-phase a.c., or d.c., or 3 or 4 cables three-phase)		Trefoil (3 cables, three-phase)	
	Current carrying capacity	Volt drop per ampere per metre	Current carrying capacity	Volt drop per ampere per metre	Current carrying capacity	Volt drop per ampere per metre	Current carrying capacity	Volt drop per ampere per metre	Current carrying capacity	Volt drop per ampere per metre	Current carrying capacity	Volt drop per ampere per metre
1	2	3	4	5	6	7	8	9	10	11	12	13
mm²	A	mV	A	mV	A	mV	A	mV	A	mV	A	mV
1.0	15	54	12	46	18	54	17	46	–	–	–	–
1.5	19	34	15	29	24	34	22	29	–	–	–	–
2.5	29	19	24	16	35	19	31	17	–	–	–	–
4	37	12	33	10	46	12	41	10	–	–	–	–
6	48	7.7	42	6.6	59	7.7	55	6.6	–	–	–	–
10	67	4.6	59	4.0	81	4.6	73	4.0	–	–	–	–
16	89	2.9	80	2.5	109	2.9	98	2.5	–	–	–	–
25	114	1.8	108	1.6	140	1.8	128	1.6	–	–	–	–
35	140	1.3	134	1.2	177	1.3	158	1.2	–	–	–	–
		a.c. d.c.				a.c. d.c.				1ph a.c. – d.c. 3ph a.c.		
50	170	1.0 0.96	145	0.89	205	0.97 0.96	185	0.86	235	0.99 0.96 0.88	210	0.84
70	215	0.75 0.67	185	0.65	260	0.68 0.67	235	0.62	295	0.71 0.67 0.65	265	0.59
95	265	0.58 0.48	225	0.51	320	0.51 0.48	285	0.47	360	0.54 0.48 0.51	325	0.44
120	310	0.51 0.38	260	0.44	370	0.41 0.38	335	0.40	425	0.46 0.38 0.44	380	0.36
150	350	0.46 0.31	300	0.40	420	0.35 0.31	380	0.35	480	0.40 0.31 0.40	425	0.30
185	–	– –	–	–	480	0.30 0.25	435	0.31	560	0.36 0.25 0.37	490	0.26
240	–	– –	–	–	570	0.25 0.19	520	0.27	660	0.36 0.19 0.39	590	0.22
300	–	– –	–	–	660	0.22 0.15	600	0.25	770	0.33 0.15 0.35	690	0.19
400	–	– –	–	–	770	0.20 0.12	700	0.24	900	0.30 0.12 0.33	790	0.18
500	–	– –	–	–	890	0.19 0.093	800	0.23	1070	0.28 0.093 0.32	910	0.16
630	–	– –	–	–	1050	0.18 0.071	950	0.22	1250	0.26 0.071 0.30	1060	0.15

†For installation method C, the tabulated values are applicable only to the range up to and including 35mm².

NOTE – WHERE THE CONDUCTOR IS TO BE PROTECTED BY A SEMI-ENCLOSED FUSE TO BS 3036, SEE ITEM 4(ii) OF THE PREFACE TO THIS APPENDIX.

CORRECTION FACTORS

FOR AMBIENT TEMPERATURE

Ambient temperature	25°C	35°C	40°C	45°C	50°C	55°C	60°C	65°C	70°C	75°C
Correction factor	1.05	0.94	0.89	0.83	0.77	0.71	0.64	0.57	0.49	0.39

FOR GROUPING

See Table 9B except that the factors for 32, 36 and 40 conductors do not apply to three-phase.

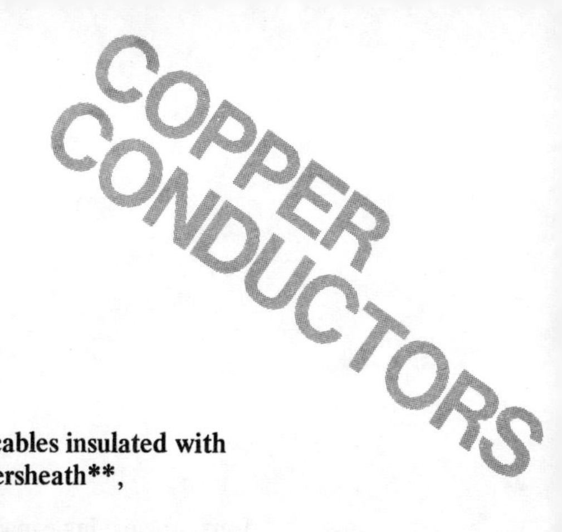

TABLE 9G1

Current-carrying capacities and associated volt drops for single-core cables insulated with impregnated paper (BS 6480), lead sheathed, with p.v.c. oversheath, non-armoured (copper conductors)**

Conductor operating temperature: 80°C

Nominal cross-sectional area of conductor	Installation methods E and F† of Table 9A ('Clipped direct')					Installation method J of Table 9A ('Defined conditions')							
	2 cables, single-phase a.c., or d.c.			3 or 4 cables, three-phase a.c.		2 cables flat or vertical, single-phase a.c., or d.c.			3 or 4 cables flat or vertical, three-phase a.c.		3 cables in trefoil, three-phase a.c.	
	Current carrying capacity	Volt drop per ampere per metre a.c.	Volt drop per ampere per metre d.c.	Current carrying capacity	Volt drop per ampere per metre	Current carrying capacity	Volt drop per ampere per metre a.c.	Volt drop per ampere per metre d.c.	Current carrying capacity	Volt drop per ampere per metre	Current carrying capacity	Volt drop per ampere per metre
1	2	3	4	5	6	7	8	9	10	11	12	13
mm²	A	mV	mV	A	mV	A	mV	mV	A	mV	A	mV
50	190	0.93	0.93	180	0.82	230	0.94	0.93	220	0.84	205	0.81
70	240	0.64	0.64	230	0.61	290	0.68	0.64	280	0.61	255	0.58
95	300	0.48	0.47	285	0.48	355	0.56	0.47	345	0.47	315	0.43
120	350	0.40	0.37	340	0.39	415	0.48	0.37	405	0.41	370	0.35
150	405	0.33	0.30	390	0.35	475	0.42	0.30	460	0.38	420	0.30
185	470	0.29	0.24	450	0.31	550	0.36	0.24	535	0.35	485	0.26
240	580	0.25	0.18	575	0.28	675	0.38	0.18	670	0.33	580	0.22
300	670	0.22	0.14	660	0.26	770	0.36	0.14	760	0.32	670	0.20
400	775	0.20	0.11	765	0.23	890	0.31	0.11	870	0.30	775	0.18
500	895	0.18	0.09	870	0.21	1000	0.29	0.09	975	0.28	885	0.17
630	1030	0.17	0.07	990	0.18	1150	0.27	0.07	1100	0.26	1020	0.16
800	1160	0.16	0.05	1100	0.17	1290	0.26	0.05	1220	0.24	1150	0.15
1000	1280	0.16	0.04	1190	0.15	1420	0.25	0.04	1330	0.22	1270	0.15

†For installation method C, see ERA Report 69–30. For cables in ducts in the floor of a building, the ERA ratings must be adjusted by the appropriate factor for ambient temperature.

NOTE – WHERE THE CONDUCTOR IS TO BE PROTECTED BY A SEMI-ENCLOSED FUSE TO BS 3036, SEE ITEM 4(ii) OF THE PREFACE TO THIS APPENDIX.

CORRECTION FACTORS

FOR AMBIENT TEMPERATURE

Ambient temperature	25°C	35°C	40°C	45°C	50°C	55°C	60°C	65°C	70°C	75°C
Correction factor	1.05	0.95	0.89	0.82	0.75	0.68	0.61	0.53	0.43	0.30

**FOR UNSERVED LEAD-SHEATHED CABLES

Cross sectional area of conductor (mm²)	50 to 185	240 to 500	630 to 1000
2 or 3 cables, flat formation	0.95	1.00	1.01
3 cables, trefoil formation	0.93	0.94	0.96

FOR GROUPING

See Table 9B except that the factors for 32, 36 and 40 conductors do not apply to 3-phase circuits.

TABLE 9G2

Current-carrying capacities and associated volt drops for twin and multicore cables insulated with impregnated paper (BS 6480), lead-sheathed or aluminium-sheathed, armoured or non-armoured, with or without serving (copper conductors).

Conductor operating temperature: 80°C

Nominal cross-sectional area of conductor*	Installation methods E and F† of Table 9A ('Clipped direct')					Installation method K of Table 9A ('Defined conditions')				
	One twin cable, single-phase a.c., or d.c.			One three- or four-core cable, three-phase		One twin cable, single-phase a.c., or d.c.			One three-or four-core cable, three-phase	
	Current carrying capacity	Volt drop per ampere per metre		Current carrying capacity	Volt drop per ampere per metre	Current carrying capacity	Volt drop per ampere per metre		Current carrying capacity	Volt drop per ampere per metre
		a.c.	d.c.				a.c.	d.c.		
1	2	3	4	5	6	7	8	9	10	11
mm²	A	mV	mV	A	mV	A	mV	mV	A	mV
50	165	0.95	0.95	145	0.82	200	0.95	0.95	170	0.82
70	205	0.66	0.66	180	0.58	250	0.66	0.66	215	0.58
95	255	0.49	0.47	225	0.43	305	0.49	0.47	265	0.43
120	295	0.40	0.36	260	0.35	355	0.40	0.36	305	0.35
150	335	0.33	0.30	300	0.28	405	0.33	0.30	350	0.28
185	390	0.28	0.24	345	0.24	465	0.28	0.24	405	0.24
240	460	0.24	0.19	410	0.20	555	0.24	0.19	480	0.20
300	525	0.21	0.15	470	0.18	635	0.21	0.15	550	0.18
400	610	0.20	0.12	545	0.17	735	0.20	0.12	640	0.17

†For installation method C, see ERA Report 69–30. For cables in ducts in the floor of a building, the ERA ratings must be adjusted by the appropriate factor for ambient temperature.
*Stranded or solid conductors.

NOTE – WHERE THE CONDUCTOR IS TO BE PROTECTED BY A SEMI-ENCLOSED FUSE TO BS 3036, SEE ITEM 4(ii) OF THE PREFACE TO THIS APPENDIX.

CORRECTION FACTORS

FOR AMBIENT TEMPERATURE

Ambient temperature	25°C	35°C	40°C	45°C	50°C	55°C	60°C	65°C	70°C	75°C
Correction factor	1.05	0.95	0.89	0.82	0.75	0.68	0.61	0.53	0.43	0.3

FOR GROUPING

See Table 9B.

TABLE 9H1

Current-carrying capacities and associated volt drops for 60°C rubber-insulated flexible cables, other than flexible cords (BS 6007)

Conductor operating temperature: 60°C

Nominal cross-sectional area of conductor	Maximum diameter of wires forming conductor	Current-carrying capacity		Volt drop per ampere per metre		
		d.c. or single-phase a.c. (one twin cable, with or without earth-continuity conductor, or two single-core cables bunched)	Three-phase a.c. (one three-, four- or five-core cable)	d.c.	Single-phase a.c.	Three-phase a.c.
1	2	3	4	5	6	7
mm^2	mm	A	A	mV	mV	mV
4	0.31	33	28	11.0	11.0	9.7
6	0.31	42	36	7.3	7.3	6.6
10	0.41	57	49	4.2	4.2	3.8
16	0.41	76	66	2.7	2.7	2.4
25	0.41	100	85	1.7	1.7	1.6
35	0.41	120	100	1.2	1.24	1.1
50	0.41	150	130	0.85	0.88	0.79
70	0.51	180	160	0.60	0.64	0.56
95	0.51	220	190	0.45	0.50	0.43
120	0.51	250	220	0.35	0.41	0.35
150	0.51	290	250	0.29	0.34	0.29
185	0.51	330	285	0.23	0.30	0.25
240	0.51	380	330	0.18	0.25	0.21
300	0.51	435	385	0.14	0.23	0.18
400	0.51	520	–	0.11	0.21	–
500	0.61	590	–	0.086	0.19	–
630	0.61	680	–	0.067	0.18	–

NOTES: 1 – **WHERE THE CONDUCTOR IS TO BE PROTECTED BY A SEMI-ENCLOSED FUSE TO BS 3036, SEE ITEM 4(ii) OF THE PREFACE TO THIS APPENDIX.**

2 – The tabulated current-carrying capacities are not applicable to flexible cables wound on drums. The current-carrying capacity of a cable on a drum depends on the type of drum and may be less than one-half of the corresponding current-carrying capacity stated in the table.

CORRECTION FACTOR FOR AMBIENT TEMPERATURE

Ambient temperature	25°C	35°C	40°C	45°C	50°C	55°C
Correction factor	1.04	0.91	0.82	0.71	0.57	0.41

TABLE 9H2

Current-carrying capacities and associated volt drops for 85°C or 150°C rubber-insulated flexible cables

(BS 6007)

Conductor operating temperature: 75°C†

Nominal cross-sectional area of conductor	Maximum diameter of wires forming conductor	Current-carrying capacity		Volt drop per ampere per metre		
		d.c. or single-phase a.c. (one twin cable, with or without earth-continuity conductor, or two single-core cables bunched)	Three-phase a.c. (one three-, four- or five-core cable)	d.c.	Single-phase a.c.	Three-phase a.c.
1	2	3	4	5	6	7
mm²	mm	A	A	mV	mV	mV
4	0.31	40	34	13.0	13.0	11.5
6	0.31	51	44	7.9	7.9	7.2
10	0.41	70	60	4.6	4.6	4.2
16	0.41	93	81	2.9	2.9	2.6
25	0.41	120	105	1.9	1.9	1.7
35	0.41	145	125	1.3	1.3	1.2
50	0.41	185	160	0.93	0.95	0.85
70	0.51	225	195	0.65	0.68	0.61
95	0.51	270	235	0.49	0.53	0.47
120	0.51	305	270	0.38	0.43	0.38
150	0.51	355	305	0.31	0.36	0.31
185	0.51	405	350	0.26	0.32	0.27
240	0.51	465	405	0.20	0.27	0.22
300	0.51	530	470	0.16	0.24	0.19
400	0.51	630	–	0.12	0.21	–
500	0.61	720	–	0.10	0.20	–
630	0.61	830	–	0.08	0.19	–

†For 150°C cables, where the correction factors for ambient temperature are used, the conductor operating temperature may be up to 150°C.

NOTES: 1 – **WHERE THE CONDUCTOR IS TO BE PROTECTED BY A SEMI-ENCLOSED FUSE TO BS 3036, SEE ITEM 4(ii) OF THE PREFACE TO THIS APPENDIX.**

2 – The tabulated current-carrying capacities are not applicable to flexible cables wound on drums. The current-carrying capacity of a cable on a drum depends on the type of drum and may be less than one-half of the corresponding capacity stated in the table.

CORRECTION FACTOR FOR AMBIENT TEMPERATURE

85°C rubber-insulated cables

Ambient temperature	35°C	40°C	45°C	50°C	55°C	60°C	65°C	70°C
Correction factor	0.93	0.86	0.80	0.72	0.63	0.54	0.44	0.31

150°C rubber-insulated cables

Ambient temperature	35°C to 95°C	100°C	105°C	110°C	115°C	120°C	125°C	130°C	135°C	140°C
Correction factor	1.0	0.94	0.88	0.82	0.77	0.71	0.64	0.56	0.48	0.39

NOTE – BS 6007 does not include 150°C rubber-insulated cables above 16mm² nominal cross-sectional area.

TABLE 9H3

Current-carrying capacities and associated volt drops and masses supportable, for flexible cords to BS 6500

Nominal cross-sectional area of conductor	Maximum diameter of wires forming conductor	Current-carrying capacity, d.c. or single-phase a.c., or three-phase a.c.	Volt drop per ampere per metre		Maximum mass supportable by twin flexible cord (see Regulation 523–32)
			d.c. or single-phase a.c.	Three-phase a.c.	
1	2	3	4	5	6
mm²	mm	A	mV	mV	kg
0.5	0.21	3	83	72	2
0.75	0.21	6	56	48	3
1.0	0.21	10	43	37	5
1.25	0.26	13	35	29	5
1.5	0.26	15	31	26	5
2.5	0.26	20	18	16	5
4	0.31	25	11	9.6	5

CORRECTION FACTOR FOR AMBIENT TEMPERATURE

60°C rubber and p.v.c. cords

Ambient temperature	35°C	40°C	45°C	50°C	55°C
Correction factor	0.96	0.92	0.87	0.71	0.50

85°C rubber cords having a h.o.f.r. sheath or a heat-resisting p.v.c. sheath

Ambient temperature	35°C to 50°C	55°C	60°C	65°C	70°C
Correction factor	1.0	0.96	0.83	0.67	0.47

150°C rubber cords

Ambient temperature	35°C to 120°C	125°C	130°C	135°C	140°C	145°C
Correction factor	1.0	0.96	0.85	0.74	0.60	0.42

Glass-fibre cords

Ambient temperature	35°C to 150°C	155°C	160°C	165°C	170°C	175°C
Correction factor	1.0	0.92	0.82	0.71	0.57	0.40

TABLE 9J1

Current-carrying capacities and associated volt drops for light-duty mineral-insulated cables (copper conductors and sheath) (BS 6207, Part 1) exposed to touch or having an overall covering of p.v.c.

Sheath operating temperature: 70°C

Nominal cross-sectional area of conductor	Two single-core cables, single-phase a.c., or d.c.		Three or four single-core cables, three-phase a.c.		One twin cable, single-phase a.c., or d.c.		One three-core cable, three-phase a.c.		One four-core cable, three-phase a.c.		One seven-core cable, all cores fully loaded		
	Current carrying capacity	Volt drop per ampere per metre	Current carrying capacity	Volt drop per ampere per metre	Current carrying capacity	Volt drop per ampere per metre	Current carrying capacity	Volt drop per ampere per metre	Current carrying capacity	Volt drop per ampere per metre	Current carrying capacity	Volt drop per ampere per metre	
												1-ph. a.c., or d.c.	3-ph. a.c.
1	2	3	4	5	6	7	8	9	10	11	12	13	14
mm²	A	mV	A	mV	A	mV	A	mV	A	mV	A	mV	mV
1.0	22	42	18	36	17	42	14	36	15	36	10	42	36
1.5	27	28	23	24	22	28	18	24	19	24	13	28	24
2.5	36	17	31	14	29	17	24	14	25	14	17	17	14
4	46	10	41	9.0	38	10	33	9.0	33	9.0	–	–	–
6	59	6.9	52	6.0	49	6.9	–	–	–	–	–	–	–
10	80	4.2	70	3.6	–	–	–	–	–	–	–	–	–

CORRECTION FACTORS

FOR AMBIENT TEMPERATURE

Ambient temperature	25°C	30°C	40°C	50°C	60°C
Correction factor for cables exposed to touch	1.06	1.0	0.85	0.68	0.46
Correction factor for cables having overall p.v.c. covering	1.16	1.1	0.94	0.75	0.51

FOR GROUPING

See Table 9B except that the factors for 32, 36 and 40 conductors do not apply to single-core, 3-phase.

TABLE 9J2

Current-carrying capacities and associated volt drops for light duty mineral-insulated cables (copper conductors and sheath) (BS 6207, Part 1) having the sheath bare and not exposed to touch and not in contact with combustible materials

Sheath operating temperature: 90°C†

Nominal cross-sectional area of conductor	Two single-core cables, single-phase a.c., or d.c., or three or four single-core cables, three-phase a.c.			One twin cable, single-phase a.c., or d.c.		One three-core cable, three-phase a.c.		One four-core cable, three-phase a.c.		One seven-core cable, all cores fully loaded		
	Current carrying capacity	Volt drop per ampere per metre: Single-phase a.c., or d.c.	Volt drop per ampere per metre: Three-phase a.c.	Current carrying capacity	Volt drop per ampere per metre	Current carrying capacity	Volt drop per ampere per metre	Current carrying capacity	Volt drop per ampere per metre	Current carrying capacity	Volt drop per ampere per metre: Single-phase a.c., or d.c.	Volt drop per ampere per metre: Three-phase a.c.
1	2	3	4	5	6	7	8	9	10	11	12	13
mm^2	A	mV	mV	A	mV	A	mV	A	mV	A	mV	mV
1.0	25	45	39	21	45	18	39	18	39	13	45	39
1.5	32	30	26	26	30	22	26	23	26	15	30	26
2.5	43	18	15	36	18	30	15	31	15	22	18	15
4	56	11	9.6	47	11	40	9.6	42	9.6	–	–	–
6	73	7.4	6.4	60	7.4	–	–	–	–	–	–	–
10	98	4.5	3.9	–	–	–	–	–	–	–	–	–

†Where the correction factors for cables having 135°C terminations and for temperatures above 30°C for cables having 105°C terminations are used, the sheath temperature will exceed 90°C and may reach the temperature limit imposed by the termination. For higher temperature working conditions, the manufacturer should be consulted.

CORRECTION FACTORS

FOR AMBIENT TEMPERATURE

Ambient temperature	25°C	30°C	40°C	50°C	60°C	70°C	80°C	90°C	100°C	110°C	120°C	130°C
Correction factor for cables having 105°C terminations	1.02	1.0	0.96	0.91	0.84	0.73	0.67	0.45	0.24	–	–	–
Correction factor for cables having 135°C terminations	1.1	1.08	1.04	1.0	0.96	0.92	0.88	0.82	0.71	0.59	0.44	0.24

FOR GROUPING

No correction factor for grouping need be applied.

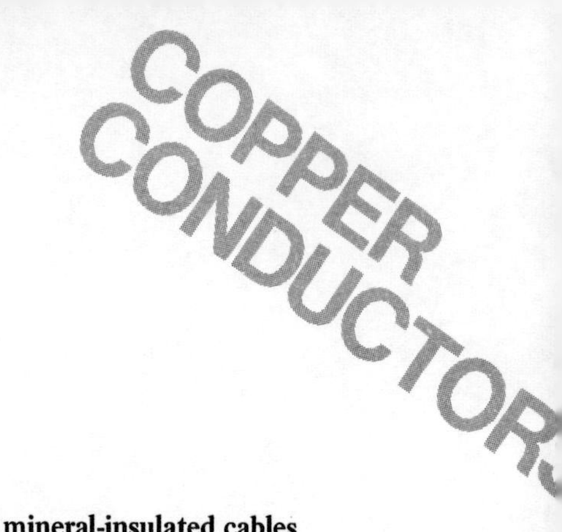

TABLE 9J3

Current-carrying capacities and associated volt drops for heavy duty mineral-insulated cables (copper conductors and sheath) (BS 6207, Part 1) exposed to touch or having an overall covering of p.v.c.

Sheath operating temperature: 70°C

Nominal cross-sectional area of conductor	Two single-core cables, single-phase a.c., or d.c.		Three or four single-core cables, three-phase a.c.		One twin cable, single-phase a.c., or d.c.		One three-core cable, three-phase a.c.		One four-core cable, three-phase a.c.		One seven-core cable, all cores fully loaded		
	Current carrying capacity	Volt drop per ampere per metre	Current carrying capacity	Volt drop per ampere per metre	Current carrying capacity	Volt drop per ampere per metre	Current carrying capacity	Volt drop per ampere per metre	Current carrying capacity	Volt drop per ampere per metre	Current carrying capacity	Volt drop per ampere per metre	
												1-ph. a.c., or d.c.	3-ph. a.c.
1	2	3	4	5	6	7	8	9	10	11	12	13	14
mm^2	A	mV	A	mV	A	mV	A	mV	A	mV	A	mV	mV
1.0	23	42	20	36	19	42	16	36	16	36	11	42	36
1.5	29	28	26	24	24	28	20	24	20	24	14	28	24
2.5	39	17	34	14	32	17	26	14	27	14	19	17	14
4	50	10	44	9.0	41	10	34	9.0	35	9.0	24	10	9.0
6	63	6.9	56	6.0	53	6.9	44	6.0	45	6.0	–	–	–
10	85	4.2	75	3.6	71	4.2	59	3.6	61	3.6	–	–	–
16	110	2.6	99	2.3	94	2.6	78	2.3	81	2.3	–	–	–
25	150	1.7	130	1.4	124	1.7	105	1.4	110	1.4	–	–	–
35	180	1.2	160	1.0	–	–	–	–	–	–	–	–	–
50	225	0.83	200	0.72	–	–	–	–	–	–	–	–	–
70	275	0.59	240	0.51	–	–	–	–	–	–	–	–	–
95	330	0.44	290	0.38	–	–	–	–	–	–	–	–	–
120	380	0.35	335	0.30	–	–	–	–	–	–	–	–	–
150	440	0.28	385	0.24	–	–	–	–	–	–	–	–	–

CORRECTION FACTORS

FOR AMBIENT TEMPERATURE

Ambient temperature	25°C	30°C	40°C	50°C	60°C
Correction factor for cables exposed to touch	1.06	1.0	0.85	0.68	0.46
Correction factor for cables having overall p.v.c. covering	1.16	1.1	0.94	0.75	0.51

FOR GROUPING

See Table 9B except that the factors for 32, 36 and 40 conductors do not apply to single-core, 3-phase.

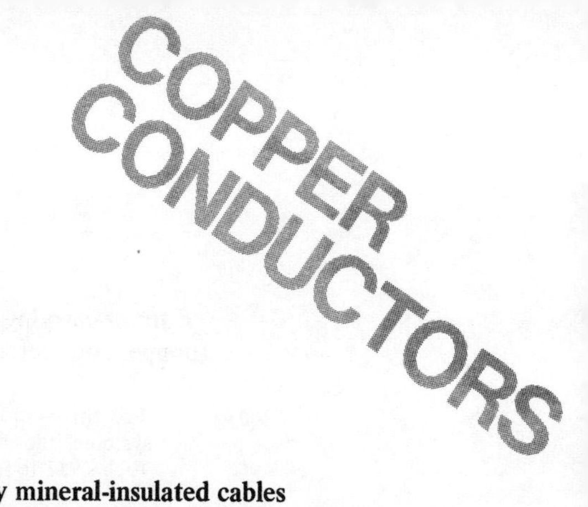

TABLE 9J4

Current-carrying capacities and associated volt drops for heavy duty mineral-insulated cables (copper conductors and sheath) (BS 6207, Part 1) having the sheath bare and not exposed to touch and not in contact with combustible materials

Sheath operating temperature: 90°C†

Nominal cross-sectional area of conductor	Two single-core cables, single-phase a.c., or d.c., or three or four single-core cables, three-phase a.c.			One twin cable, single-phase a.c., or d.c.		One three-core cable, three-phase a.c.		One four-core cable, three-phase a.c.		One seven-core cable, all cores fully loaded		
	Current carrying capacity	Volt drop per ampere per metre		Current carrying capacity	Volt drop per ampere per metre	Current carrying capacity	Volt drop per ampere per metre	Current carrying capacity	Volt drop per ampere per metre	Current carrying capacity	Volt drop per ampere per metre	
		Single-phase a.c., or d.c.	Three phase a.c.								Single-phase a.c., or d.c.	Three-phase a.c.
1	2	3	4	5	6	7	8	9	10	11	12	13
mm²	A	mV	mV	A	mV	A	mV	A	mV	A	mV	mV
1.0	29	45	39	23	45	19	39	20	39	14	45	39
1.5	36	30	26	29	30	24	26	25	26	17	30	26
2.5	47	18	15	39	18	32	15	33	15	23	18	15
4	62	11	9.6	51	11	42	9.6	44	9.6	30	11	9.6
6	77	7.4	6.4	65	7.4	54	6.4	55	6.4	–	–	–
10	105	4.5	3.9	87	4.5	73	3.9	75	3.9	–	–	–
16	140	2.8	2.4	115	2.8	98	2.4	100	2.4	–	–	–
25	180	1.8	1.5	155	1.8	125	1.5	130	1.5	–	–	–
35	220	1.3	1.1	–	–	–	–	–	–	–	–	–
50	275	0.89	0.77	–	–	–	–	–	–	–	–	–
70	335	0.64	0.55	–	–	–	–	–	–	–	–	–
95	405	0.47	0.41	–	–	–	–	–	–	–	–	–
120	470	0.37	0.32	–	–	–	–	–	–	–	–	–
150	540	0.30	0.26	–	–	–	–	–	–	–	–	–

†Where the correction factors for cables having 135°C terminations and for temperatures above 30°C for cables having 105°C terminations are used, the sheath temperature will exceed 90°C and may reach the temperature limit imposed by the termination. For higher temperature working conditions, the manufacturer should be consulted.

CORRECTION FACTORS

FOR AMBIENT TEMPERATURE

Ambient temperature	25°C	30°C	40°C	50°C	60°C	70°C	80°C	90°C	100°C	110°C	120°C	130°C
Correction factor for cables having 105°C terminations	1.02	1.0	0.96	0.91	0.84	0.73	0.67	0.45	0.24	–	–	–
Correction factor for cables having 135°C terminations	1.1	1.08	1.04	1.0	0.96	0.92	0.88	0.82	0.71	0.59	0.44	0.24

FOR GROUPING

No factor need be applied.

TABLE 9J5

Current-carrying capacities and associated volt drops for mineral-insulated cables (copper conductors and sheath) (BS 6207 Part 1) used as earthed-concentric wiring

NOTE – For three-core mineral-insulated cables carrying balanced three-phase a.c. in earthed-concentric systems, the current-carrying capacities and values of volt drop stated for three-core cables in Tables 9J1 to 9J4 as appropriate, are applicable.

Sheath operating temperatures:
Exposed to touch, 70°C;
Bare and not exposed, 90°C††

Nominal cross-sectional area of conductor	Single-core cables exposed to touch or having an overall covering of p.v.c.		Single-core cables having the sheath bare and not exposed to touch and not in contact with combustible materials	
	Current carrying capacity	Volt drop per ampere per metre	Current carrying capacity	Volt drop per ampere per metre
1	2	3	4	5
mm^2	A	mV	A	mV
†1.0	18	29	22	31
†1.5	21	21	26	22
†2.5	27	14	33	15
†4	34	10	42	10
*1.0	21	25	26	27
*1.5	26	18	32	19
*2.5	33	12	40	13
*4	41	7.8	50	8.5
*6	49	5.9	60	6.3

† Light duty cables
*Heavy duty cables
††Where the correction factors for cables having 135°C terminations and for temperatures above 30°C for cables having 105°C terminations are used, the sheath temperature will exceed 90°C and may reach the temperature limit imposed by the termination. For higher temperature working conditions, the manufacturer should be consulted.

CORRECTION FACTORS

FOR AMBIENT TEMPERATURE

Ambient temperature
25°C 30°C 40°C 50°C 60°C 70°C 80°C 90°C 100°C 110°C 120°C 130°C

Correction factor for cables exposed to touch (applicable to ratings in Col. 2)
1.06 1.0 0.85 0.68 0.46 – – – – – – –

Correction factor for cables having overall p.v.c. covering (applicable to ratings in Col. 2)
1.16 1.1 0.94 0.75 0.51 – – – – – – –

Correction factor for cables having 105°C terminations (applicable to ratings in Col. 4)
1.02 1.0 0.96 0.91 0.84 0.73 0.67 0.45 0.24 – – – –

Correction factor for cables having 135°C terminations (applicable to ratings in Col. 4)
1.1 1.08 1.04 1.0 0.96 0.92 0.88 0.82 0.71 0.59 0.44 0.24

FOR GROUPING

For cables exposed to touch or in contact with combustible material or having an overall covering of p.v.c. use Table 9B.

For cables having the sheath bare and not exposed to touch no correction factor for grouping need be applied.

TABLE 9J6

Current-carrying capacities and associated volt drops for light-duty mineral-insulated cables (copper conductors and aluminium sheath) (BS 6207 Part 1) having an overall covering of p.v.c.

Sheath operating temperature: 70°C

Nominal cross-sectional area of conductor	One twin cable, single-phase a.c., or d.c.		One three-core cable, three-phase a.c.		One four-core cable, three-phase a.c.		One seven-core cable, all cores fully loaded		
	Current carrying capacity	Volt drop per ampere per metre	Current carrying capacity	Volt drop per ampere per metre	Current carrying capacity	Volt drop per ampere per metre	Current carrying capacity	Volt drop per ampere per metre	
								1-phase a.c., or d.c.	3-phase a.c.
1	2	3	4	5	6	7	8	9	10
mm²	A	mV	A	mV	A	mV	A	mV	mV
1.0	20	42	16	36	17	36	12	42	36
1.5	25	28	21	24	21	24	15	28	24
2.5	33	17	28	14	29	14	20	17	14
4.0	44	10	37	9.0	38	9.0	–	–	–
6.0	56	6.9	47	6.0	48	6.0	–	–	–
10	76	4.2	64	3.6	–	–	–	–	–

CORRECTION FACTORS

FOR AMBIENT TEMPERATURE

Ambient temperature	25°C	30°C	40°C	50°C	60°C
Correction factor	1.06	1.0	0.85	0.68	0.46

FOR GROUPING

See Table 9B except that the factors for 32, 36 and 40 conductors do not apply to single-core, 3-phase.

TABLE 9K1

Current-carrying capacities and associated voltage drops for single-core p.v.c.-insulated cables, non-armoured, with sheath (aluminium conductors)

BS 6346

Conductor operating temperature: 70°C

Cross sectional area of conductor	Installation methods A to C† of Table 9A ('Enclosed')					Installation methods E to H of Table 9A ('Clipped direct')					Installation method J of Table 9A ('Defined conditions')					
	2 cables, single-phase a.c., or d.c.			3 or 4 cables, three-phase a.c.		2 cables, single-phase a.c., or d.c.			3 or 4 cables, three-phase a.c.		Flat or vertical (2 cables, single-phase a.c., or d.c., or 3 or 4 cables three-phase)				Trefoil (3 cables three-phase)	
	Current carrying capacity	Volt drop per ampere per metre a.c.	Volt drop per ampere per metre d.c.	Current carrying capacity	Volt drop per ampere per metre	Current carrying capacity	Volt drop per ampere per metre a.c.	Volt drop per ampere per metre d.c.	Current carrying capacity	Volt drop per ampere per metre	Current carrying capacity	Volt drop per ampere per metre 1ph	Volt drop per ampere per metre d.c.	Volt drop per ampere per metre 3ph	Current carrying capacity	Volt drop per ampere per metre
1	2	3	4	5	6	7	8	9	10	11	12	13	14	15	16	17
mm²	A	mV	mV	A	mV	A	mV	mV	A	mV	A	mV	mV	mV	A	mV
16	60	4.5	4.5	52	3.9	72	4.5	4.5	65	3.9	–	–	–	–	–	–
25	78	2.9	2.8	67	2.5	94	2.8	2.8	85	2.5	–	–	–	–	–	–
35	96	2.1	2.0	83	1.8	115	2.1	2.0	105	1.8	–	–	–	–	–	–
50	120	1.6	1.5	100	1.4	143	1.5	1.5	123	1.3	155	1.5	1.5	1.34	140	1.3
70	150	1.2	1.0	125	1.0	181	1.1	1.0	156	0.93	190	1.1	1.0	0.95	170	0.90
95	175	0.93	0.75	150	0.80	223	0.77	0.75	193	0.69	235	0.80	0.75	0.72	205	0.67
120	205	0.80	0.60	175	0.70	261	0.62	0.60	225	0.56	275	0.65	0.60	0.60	235	0.54
150	235	0.73	0.49	200	0.64	298	0.51	0.49	259	0.48	320	0.55	0.49	0.51	270	0.45
185	–	–	–	–	–	345	0.42	0.39	290	0.40	370	0.46	0.39	0.45	310	0.37
240	–	–	–	–	–	411	0.34	0.29	361	0.34	440	0.43	0.29	0.43	370	0.30
300	–	–	–	–	–	476	0.29	0.23	419	0.30	510	0.38	0.23	0.39	435	0.25
380	–	–	–	–	–	554	0.26	0.19	465	0.28	584	0.35	0.19	0.37	490	0.22
480	–	–	–	–	–	643	0.23	0.15	541	0.26	677	0.32	0.15	0.34	570	0.20
600	–	–	–	–	–	737	0.21	0.12	616	0.24	776	0.30	0.12	0.33	648	0.18

†For installation method C, the tabulated values are applicable only to the range up to and including 35mm². For larger sizes in this installation method, see ERA Report 69–30. For cables in ducts in the floor of a building, the ERA ratings must be adjusted by the appropriate factor for ambient temperature.

NOTE – WHERE THE CONDUCTOR IS TO BE PROTECTED BY A SEMI-ENCLOSED FUSE TO BS 3036, SEE ITEM 4(ii) OF THE PREFACE TO THIS APPENDIX.

CORRECTION FACTORS

FOR AMBIENT TEMPERATURE

Ambient temperature	25°C	35°C	40°C	45°C	50°C	55°C	60°C	65°C
Correction factor	1.06	0.94	0.87	0.79	0.71	0.61	0.50	0.35

FOR GROUPING

See Table 9B, except that the factors for 32, 36 and 40 conductors do not apply to three-phase.

TABLE 9K2

Current-carrying capacities and associated voltage drops for twin and multicore p.v.c.-insulated cables, non-armoured (aluminium conductors)

BS 6346

Conductor operating temperature: 70°C

	Installation methods E to H† of Table 9A ('Clipped direct')				Installation method K of Table 9A ('Defined conditions')			
Conductor cross-sectional area	One twin cable single-phase a.c., or d.c.		One three- or four-core cable, three-phase		One twin cable, single-phase a.c., or d.c.		One three- or four-core cable, three-phase	
	Current carrying capacity	Volt drop per ampere per metre	Current carrying capacity	Volt drop per ampere per metre	Current carrying capacity	Volt drop per ampere per metre	Current carrying capacity	Volt drop per ampere per metre
1	2	3	4	5	6	7	8	9
mm^2	A	mV	A	mV	A	mV	A	mV
16	62	4.5	53	3.9	65	4.5	55	3.9
25	82	2.9	70	2.5	86	2.9	74	2.5
35	102	2.1	86	1.8	107	2.1	91	1.8
50	120	1.5	106	1.3	125	1.5	110	1.3
70	150	1.1	133	0.93	158	1.1	139	0.93
95	185	0.79	163	0.68	195	0.79	172	0.68
120	–	–	190	0.54	–	–	200	0.54
150	–	–	217	0.45	–	–	227	0.45
185	–	–	247	0.37	–	–	260	0.37
240	–	–	296	0.29	–	–	311	0.29
300	–	–	340	0.25	–	–	358	0.25

†For installation method C, see ERA Report 69–30. For cables in ducts in the floor of a building, the ERA ratings must be adjusted by the appropriate factor for ambient temperature.

NOTE – WHERE THE CONDUCTOR IS TO BE PROTECTED BY A SEMI-ENCLOSED FUSE TO BS 3036, SEE ITEM 4 (ii) OF THE PREFACE TO THIS APPENDIX.

CORRECTION FACTORS

FOR AMBIENT TEMPERATURE

Ambient temperature	25°C	35°C	40°C	45°C	50°C	55°C	60°C	65°C
Correction factor	1.06	0.94	0.87	0.79	0.71	0.61	0.50	0.35

FOR GROUPING

See Table 9B.

TABLE 9K3

Current-carrying capacities and associated voltage drops for twin and multicore armoured p.v.c.-insulated cables (aluminium conductors)

BS 6346

Conductor operating temperature: 70°C

Cross-sectional area of conductor	Installation methods E, F and G† of Table 9A ('Clipped direct')					Installation method K of Table 9A ('Defined conditions')				
	One twin cable, single-phase a.c., or d.c.			One three- or four-core cable, three-phase		One twin cable, single-phase a.c., or d.c.			One three- or four-core cable, three-phase	
	Current carrying capacity	Volt drop per ampere per metre		Current carrying capacity	Volt drop per ampere per metre	Current carrying capacity	Volt drop per ampere per metre		Current carrying capacity	Volt drop per ampere per metre
		a.c.	d.c.				a.c.	d.c.		
1	2	3	4	5	6	7	8	9	10	11
mm²	A	mV	mV	A	mV	A	mV	mV	A	mV
16	63	4.5	4.5	55	3.9	66	4.5	4.3	58	3.9
25	83	2.9	2.9	67	2.5	87	2.9	2.9	71	2.5
35	100	2.1	2.0	88	1.8	105	2.1	2.0	93	1.8
50	124	1.6	1.5	105	1.3	130	1.6	1.5	110	1.3
70	157	1.1	1.0	138	0.93	165	1.1	1.0	145	0.93
95	185	0.79	0.77	166	0.68	195	0.79	0.77	175	0.68
120	–	–	–	195	0.54	–	–	–	205	0.54
150	–	–	–	219	0.45	–	–	–	230	0.45
185	–	–	–	257	0.37	–	–	–	270	0.37
240	–	–	–	304	0.30	–	–	–	320	0.30
300	–	–	–	347	0.25	–	–	–	365	0.25

†For installation method C, see ERA Report 69-30. For cables in ducts in the floor of a building, the ERA ratings must be adjusted by the appropriate factor for ambient temperature.

NOTE – WHERE THE CONDUCTOR IS TO BE PROTECTED BY A SEMI-ENCLOSED FUSE TO BS 3036, SEE ITEM 4(ii) OF THE PREFACE TO THIS APPENDIX.

CORRECTION FACTORS

FOR AMBIENT TEMPERATURE

Ambient temperature	25°C	35°C	40°C	45°C	50°C	55°C	60°C	65°C
Correction factor	1.06	0.94	0.87	0.79	0.71	0.61	0.50	0.35

FOR GROUPING

See Table 9B.

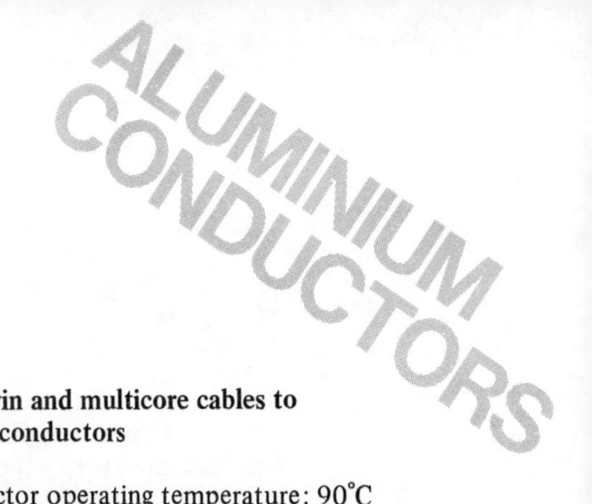

TABLE 9L1

Current-carrying capacities and associated voltage drops for armoured twin and multicore cables to BS 5467, having thermosetting insulation and aluminium conductors

Conductor operating temperature: 90°C

Conductor cross-sectional area	Installation methods E, F and G† of Table 9A ('Clipped direct')						Installation method K of Table 9A ('Defined conditions')					
	One twin cable, a.c. or d.c.			One three- or four-core cable, balanced three-phase a.c.			One twin cable, a.c. or d.c.			One three- or four-core cable, balanced three-phase a.c.		
	Current carrying capacity	Voltage drop per ampere per metre a.c.	Voltage drop per ampere per metre d.c.	Current carrying capacity	Voltage drop per ampere per metre		Current carrying capacity	Voltage drop per ampere per metre a.c.	Voltage drop per ampere per metre d.c.	Current carrying capacity	Voltage drop per ampere per metre	
1	2	3	4	5	6		7	8	9	10	11	
mm²	A	mV		A	mV		A	mV		A	mV	
16	80	4.8	4.8	69	4.2		84	4.8	4.8	73	4.2	
25	108	3.1	3.1	95	2.7		114	3.1	3.1	100	2.7	
35	130	2.2	2.2	113	1.9		137	2.2	2.2	118	1.9	
50	158	1.7	1.7	140	1.4		166	1.7	1.7	147	1.4	
70	199	1.2	1.1	176	0.99		209	1.2	1.1	185	0.99	
95	244	0.83	0.83	212	0.72		256	0.83	0.83	223	0.72	
120	–	–	–	253	0.58		–	–	–	266	0.58	
150	–	–	–	289	0.48		–	–	–	304	0.48	
185	–	–	–	334	0.39		–	–	–	351	0.39	
240	–	–	–	397	0.31		–	–	–	418	0.31	
300	–	–	–	460	0.26		–	–	–	484	0.26	

†For installation method C, see ERA Report 69-30. For cables in a duct in the floor of a building, the ERA ratings must be adjusted by the appropriate factor for ambient temperature.

NOTE – WHERE THE CONDUCTOR IS TO BE PROTECTED BY A SEMI-ENCLOSED FUSE TO BS 3036, SEE ITEM 4 (ii) OF THE PREFACE TO THIS APPENDIX.

CORRECTION FACTORS

FOR AMBIENT TEMPERATURE

Ambient temperature	25°C	35°C	40°C	45°C	50°C	55°C	60°C	65°C	70°C	75°C	80°C
Correction factor	1.04	0.96	0.91	0.87	0.82	0.76	0.71	0.65	0.58	0.50	0.41

FOR GROUPING
See Table 9B.

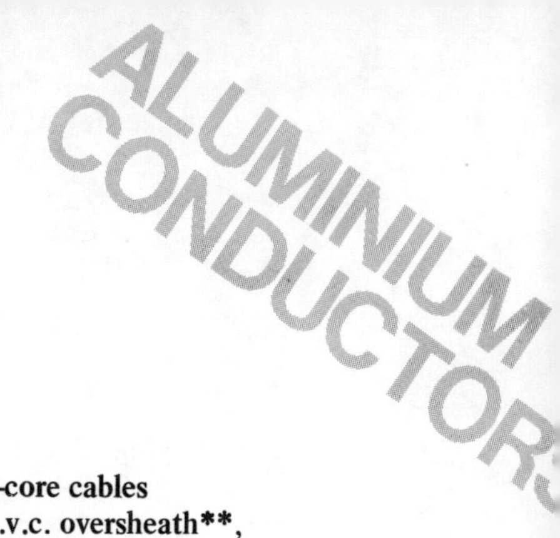

TABLE 9M1

Current-carrying capacities and associated volt drops for single-core cables insulated with impregnated paper (BS 6480), lead-sheathed, with p.v.c. oversheath, non-armoured (aluminium conductors)**

Conductor operating temperature: 80°C

Nominal cross-sectional area of conductor	Installation methods E and F† of Table 9A ('Clipped direct')					Installation method J of Table 9A ('Defined conditions')						
	2 cables, single-phase a.c., or d.c.			3 or 4 cables, three-phase a.c.		2 cables, flat, or vertical, single-phase a.c., or d.c.			3 or 4 cables flat or vertical, three-phase a.c.		3 cables in trefoil, three-phase a.c.	
	Current carrying capacity	Volt drop per ampere per metre a.c.	Volt drop per ampere per metre d.c.	Current carrying capacity	Volt drop per ampere per metre	Current carrying capacity	Volt drop per ampere per metre a.c.	Volt drop per ampere per metre d.c.	Current carrying capacity	Volt drop per ampere per metre	Current carrying capacity	Volt drop per ampere per metre
1	2	3	4	5	6	7	8	9	10	11	12	13
mm²	A	mV	mV	A	mV	A	mV	mV	A	mV	A	mV
50	145	1.6	1.6	144	1.3	175	1.6	1.6	170	1.4	155	1.3
70	185	1.1	1.1	175	0.96	225	1.1	1.1	215	1.0	200	0.92
95	225	0.8	0.77	215	0.72	275	0.82	0.77	270	0.77	245	0.68
120	265	0.64	0.61	260	0.58	325	0.68	0.61	315	0.64	285	0.55
150	300	0.52	0.51	295	0.50	370	0.58	0.51	360	0.56	325	0.46
185	355	0.44	0.40	345	0.44	430	0.50	0.40	420	0.49	380	0.38
240	435	0.36	0.33	430	0.39	530	0.47	0.33	530	0.46	455	0.32
300	495	0.30	0.24	500	0.34	610	0.41	0.24	605	0.40	525	0.27
400	590	0.26	0.19	590	0.29	710	0.34	0.19	700	0.33	615	0.23
500	670	0.23	0.15	675	0.25	810	0.30	0.15	790	0.28	710	0.20
630	790	0.21	0.12	790	0.22	940	0.26	0.12	910	0.24	830	0.18
800	910	0.19	0.09	890	0.19	1070	0.23	0.09	1030	0.21	910	0.17
1000	1020	0.17	0.07	990	0.17	1210	0.21	0.07	1140	0.18	1080	0.16

†For installation method C, see ERA Report 69-30. For cables in ducts in the floor of a building, the ERA ratings must be adjusted by the appropriate factor for ambient temperature.

NOTE – WHERE THE CONDUCTOR IS TO BE PROTECTED BY A SEMI-ENCLOSED FUSE TO BS 3036, SEE ITEM 4 (ii) OF THE PREFACE TO THIS APPENDIX.

CORRECTION FACTORS

FOR AMBIENT TEMPERATURE

Ambient temperature	25°C	35°C	40°C	45°C	50°C	55°C	60°C	65°C	70°C	75°C
Correction factor	1.05	0.95	0.89	0.82	0.75	0.68	0.61	0.53	0.43	0.30

FOR GROUPING

See Table 9B except that the factors for 32, 36 and 40 conductors do not apply to 3-phase circuits.

**FOR UNSERVED LEAD-SHEATHED CABLES

Cross-sectional area of conductor (mm²)	50 to 185	240 to 500	630 to 1000
2 or 3 cables, flat formation	0.95	1.00	1.01
3 cables, trefoil formation	0.93	0.94	0.96

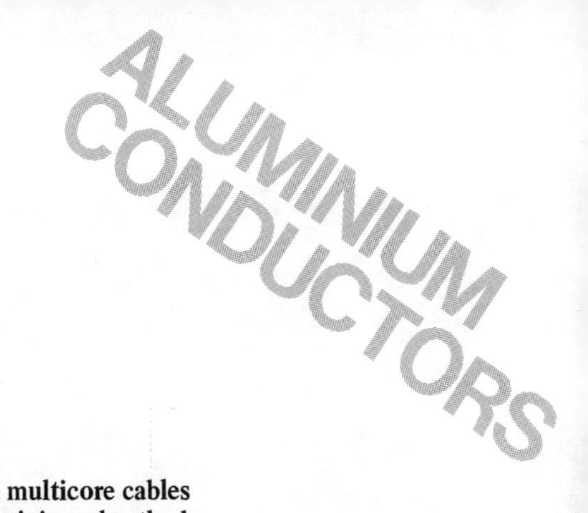

TABLE 9M2

Current-carrying capacities and associated volt drops for twin and multicore cables insulated with impregnated paper (BS 6480), lead-sheathed or aluminium-sheathed, armoured or non-armoured with or without serving (aluminium conductors)

Conductor operating temperature: 80°C

Nominal cross-sectional area of conductor	Installation methods E and F‡ of Table 9A ('Clipped direct')					Installation method K of Table 9A ('Defined conditions')				
	One twin cable, single-phase a.c., or d.c.			One three-or four-core cable three-phase		One twin cable, single-phase a.c., or d.c.			One three-or four-core cable, three-phase	
	Current carrying capacity	Volt drop per ampere per metre a.c.	Volt drop per ampere per metre d.c.	Current carrying capacity	Volt drop per ampere per metre	Current carrying capacity	Volt drop per ampere per metre a.c.	Volt drop per ampere per metre d.c.	Current carrying capacity	Volt drop per ampere per metre
1	2	3	4	5	6	7	8	9	10	11
mm²	A	mV	mV	A	mV	A	mV	mV	A	mV
50	125†	1.6	1.6	115†	1.3	145†	1.6	1.6	130†	1.3
70	165	1.1	1.1	140†	0.95	195	1.1	1.1	160†	0.95
95	205	0.81	0.79	170†	0.70	240	0.81	0.79	200†	0.70
120	235	0.64	0.63	200†	0.55	275	0.64	0.63	230†	0.55
150	265	0.53	0.53	225†	0.46	315	0.53	0.51	265†	0.46
185	310	0.44	0.41	245†	0.38	365	0.44	0.41	305†	0.38
240	370	0.36	0.31	290†	0.31	435	0.36	0.31	365†	0.31
300	425	0.30	0.25	335†	0.26	500	0.30	0.25	420†	0.26
400	495	0.26	0.19	435	0.22	585	0.26	0.19	510	0.22

†These values relate to solid conductors. They may safely be applied also to stranded conductors, but for more accurate current carrying capacities for stranded conductors ERA Report 69-30, or the Secretary of the Institution, should be consulted. Otherwise the Table relates to both solid and stranded conductors.

‡For installation method C, see ERA Report 69-30. For cables in ducts in the floor of a building, the FRA ratings must be adjusted by the appropriate factor for ambient temperature.

NOTE – WHERE THE CONDUCTOR IS TO BE PROTECTED BY A SEMI-ENCLOSED FUSE TO BS 3036, SEE ITEM 4(ii) OF THE PREFACE TO THIS APPENDIX.

CORRECTION FACTORS

FOR AMBIENT TEMPERATURE

Ambient temperature	25°C	35°C	40°C	45°C	50°C	55°C	60°C	65°C	70°C	75°C
Correction factor	1.05	0.95	0.89	0.82	0.75	0.68	0.61	0.53	0.43	0.30

FOR GROUPING

See Table 9B.

TABLE 9N1

Current-carrying capacities and associated volt drops for heavy duty mineral insulated cables (aluminium conductors and aluminium sheath) (BS 6207 Part 2) having an overall covering of p.v.c.

Sheath operating temperature: 70°C

Nominal cross-sectional area of conductor	Two single-core cables, single-phase a.c., or d.c.		Three or four single-core cables, three-phase a.c.		One twin cable, single-phase a.c., or d.c.		One three-core cable, three-phase a.c.		One four-core cable, three-phase a.c.	
	Current carrying capacity	Volt drop per ampere per metre	Current carrying capacity	Volt drop per ampere per metre	Current carrying capacity	Volt drop per ampere per metre	Current carrying capacity	Volt drop per ampere per metre	Current carrying capacity	Volt drop per ampere per metre
1	2	3	4	5	6	7	8	9	10	11
mm^2	A	mV	A	mV	A	mV	A	mV	A	mV
16	99	4.3	87	3.7	86	4.3	72	3.7	73	3.7
25	130	2.7	115	2.4	115	2.7	95	2.4	99	2.4
35	160	1.9	140	1.7	–	–	–	–	–	–
50	195	1.4	175	1.2	–	–	–	–	–	–
70	240	0.97	215	0.84	–	–	–	–	–	–
95	300	0.72	265	0.62	–	–	–	–	–	–
120	355	0.57	310	0.49	–	–	–	–	–	–
150	410	0.45	360	0.39	–	–	–	–	–	–

CORRECTION FACTORS

FOR AMBIENT TEMPERATURE

Ambient temperature	25°C	30°C	40°C	50°C	60°C
Correction factor	1.06	1.0	0.85	0.68	0.46

FOR GROUPING

See Table 9B except that the factors for 32, 36 and 40 conductors do not apply to single-core, 3-phase.

APPENDIX 10

NOTES ON THE SELECTION OF TYPES OF CABLE AND FLEXIBLE CORD FOR PARTICULAR USES AND EXTERNAL INFLUENCES

For compliance with the requirements of Chapter 52 for the selection and erection of wiring systems in relation to risks of mechanical damage and corrosion, this appendix lists in two tables types of cable and flexible cord suitable for the uses intended. These tables are not intended to be exhaustive and other limitations may be imposed by the relevant regulations, in particular those concerning maximum permissible operating temperatures.

Information is also included in this appendix on protection against corrosion of exposed metalwork of wiring systems.

TABLE 10A

Applications of cables for fixed wiring

Type of cable	Uses	Additional precautions (if any)
P.V.C.- or rubber-insulated non-sheathed	In conduits, cable ducting or trunking, but not in such conduits etc. buried underground	–
Light circular p.v.c.-insulated and sheathed	(i) General indoor use other than embedding (ii) Underground in conduit or pipes	Additional protection where exposed to severe mechanical stresses
Flat p.v.c.-insulated and sheathed	(i) General indoor use (ii) On exterior surface walls, boundary walls and the like (iii) Overhead wiring between buildings (iv) Underground in conduits or pipes	Additional protection where exposed to severe mechanical stresses
Split-concentric p.v.c.-insulated	General	–
Consac	General	–
Mineral-insulated	General	With overall p.v.c. covering where exposed to the weather or risk of corrosion, or where installed underground, or in concrete ducts
P.V.C.-insulated and armoured	General	With overall p.v.c. covering where exposed to the weather or risk of corrosion, or where installed underground, or in concrete ducts
Paper-insulated lead-sheathed	General	(i) With armouring where exposed to severe mechanical stresses or where installed underground (ii) With serving where installed in concrete ducts

NOTES: 1 – The use of cable covers (preferably conforming to BS 2484) or equivalent mechanical protection is desirable for all underground cables which might otherwise subsequently be disturbed.

2 – Cables having p.v.c. insulation or sheath should preferably not be used where the ambient temperature is consistently below 0°C. Where they are to be installed during a period of low temperature, precautions should be taken to avoid risk of mechanical damage during handling.

TABLE 10B

Applications of flexible cords

Type of flexible cord	Uses
60°C rubber-insulated braided twin and three-core	Indoors in household or commercial premises where subject only to low mechanical stresses
60°C rubber-insulated and sheathed	(i) Indoors in household or commercial premises where subject only to low mechanical stresses (ii) Occasional use outdoors
60°C rubber-insulated sheathed and screened	Portable hand-lamps on construction sites or similar applications
60°C rubber-insulated oil-resisting and flame-retardant sheath	(i) General, unless subject to severe mechanical stresses (ii) Fixed installations protected in conduit or other enclosure
85°C rubber-insulated HOFR sheathed	General, including hot situations e.g. night storage heaters and immersion heaters
85°C heat resisting p.v.c.-insulated and sheathed (to BS 6500(1969))	General, including hot situations e.g. for pendant luminaires
150°C rubber-insulated and braided	(i) At high ambient temperatures (ii) In or on luminaires
185°C glass-fibre-insulated single-core twisted twin and three-core	For internal wiring of luminaires only and then only where permitted by BS 4533
185°C glass-fibre-insulated braided circular	(i) Dry situations at high ambient and not subject to abrasion or undue flexing (ii) Wiring of luminaires
Light p.v.c.-insulated and sheathed	Indoors in household or commercial premises in dry situations, for light duty
Ordinary p.v.c.-insulated and sheathed	(i) Indoors in household or commercial premises, including damp situations, for medium duty (ii) For cooking and heating appliances where not in contact with hot parts (iii) For outdoor use other than in agricultural or industrial applications

Protection against corrosion of exposed metalwork of wiring systems

In damp situations, where metal cable sheaths and armour of cables, metal conduit and conduit fittings, metal ducting and trunking systems, and associated metal fixings, are liable to chemical or electrolytic attack by materials of a structure with which they may come in contact, it is necessary to take suitable precautions against corrosion.

Materials likely to cause such attack include –

– materials containing magnesium chloride which are used in the construction of floors and dadoes,

– plaster undercoats contaminated with corrosive salts,

– lime, cement and plaster, for example on unpainted walls,

– oak and other acidic woods,

– dissimilar metals liable to set up electrolytic action.

Application of suitable coatings before erection, or prevention of contact by separation with plastics, are recognized as effectual precautions against corrosion.

Special care is required in the choice of materials for clips and other fittings for bare aluminium-sheathed cables and for aluminium conduit, to avoid risk of local corrosion in damp situations. Examples of suitable materials for this purpose are the following:

– Porcelain,

– Plastics,

– Aluminium,

– Corrosion-resistant aluminium alloys,

– Zinc alloys complying with BS 1004,

– Iron or steel protected against corrosion by galvanizing, sherardizing, etc.

Contact between bare aluminium sheaths or aluminium conduits and any parts made of brass or other metal having a high copper content, should be especially avoided in damp situations, unless the parts are suitably plated. If such contact is unavoidable, the joint should be completely protected against ingress of moisture. Wiped joints in aluminium-sheathed cables should always be protected against moisture by a suitable paint, by an impervious tape, or by embedding in bitumen.

APPENDIX 11

NOTES ON METHODS OF SUPPORT FOR CABLES, CONDUCTORS AND WIRING SYSTEMS

This appendix describes examples of methods of support for cables, conductors and wiring systems which satisfy the relevant requirements of Section 529 of these Regulations. The use of other methods is not precluded where specified by a suitably qualified electrical engineer.

Cables generally

Items 1 to 8 below are generally applicable to supports on structures which are subject only to vibration of low severity and a low risk of mechanical impact.

1. For non-sheathed cables, installation in conduit without further fixing of the cables, precautions being taken against undue compression or other mechanical stressing of the insulation at the top of any vertical runs exceeding 5m in length.

2. For cables of any type, installation in ducting or trunking without further fixing of the cables, vertical runs not exceeding 5m in length without intermediate support.

3. For sheathed and/or armoured cables installed in accessible positions, support by clips at spacings not exceeding the appropriate value stated in Table 11A.

4. For cables of any type, resting without fixing in horizontal runs of ducts, conduits, cable ducting or trunking.

5. For sheathed and/or armoured cables in horizontal runs which are inaccessible and unlikely to be disturbed, resting without fixing on part of a building, the surface of that part being reasonably smooth.

6. For sheathed-and-armoured cables in vertical runs which are inaccessible and unlikely to be disturbed, support at the top of the run by a clip and a rounded support of a radius not less than the appropriate value stated in Table 52C of Chapter 52.

7. For sheathed cables without armour in vertical runs which are inaccessible and unlikely to be disturbed, support by the method described in Item 6 above; the length of run without intermediate support not exceeding 2m for a lead-sheathed cable or 5m for a rubber- or p.v.c.-sheathed cable.

8. For rubber- or p.v.c.-sheathed cables, installation in conduit without further fixing of the cables, any vertical runs being in conduit of suitable size and not exceeding 5m in length.

Cables in particular conditions

9. In caravans, for sheathed cables in inaccessible spaces such as ceiling, wall, and floor spaces, support at intervals not exceeding 250mm for horizontal runs and 400mm for vertical runs.

10. In caravans for horizontal runs of sheathed cables passing through floor or ceiling joists at 350 to 400mm intervals in inaccessible floor or ceiling spaces, securely bedded in thermal insulating material, no further fixing is required.

11. For flexible cords used as pendants, attachment to a ceiling rose or similar accessory by the cord grip or other method of strain relief provided in the accessory.

12. For temporary installations and installations on construction sites, supports so arranged that there is no appreciable mechanical strain on any cable termination or joint.

Overhead wiring

13. For cables sheathed with rubber or p.v.c., support by a separate catenary wire, either continuously bound up with the cable or attached thereto at intervals; the intervals not exceeding those stated in Column 2 of Table 11A.

14. Support by a catenary wire incorporated in the cable during manufacture, the spacings between supports not exceeding those stated by the manufacturer and the minimum height above ground being in accordance with Table 11B.

15. For spans without intermediate support (e.g. between buildings) of p.v.c.-insulated p.v.c.-sheathed cable, or rubber-insulated cable having an oil-resisting and flame-retardant or h.o.f.r. sheath, terminal supports so arranged that no undue strain is placed upon the conductors or insulation of the cable, adequate precautions being taken against any risk of chafing of the cable sheath, and the minimum height above ground and the length of such spans being in accordance with the appropriate values indicated in Table 11B.

16. Bare or p.v.c.-covered conductors of an overhead line for distribution between a building and a point of utilisation not attached thereto (e.g. another building) supported on insulators, the lengths of span and heights above ground having the appropriate values indicated in Table 11B or otherwise installed in accordance with the Overhead Line Regulations.*

17. For spans without intermediate support (e.g. between buildings) and which are in situations inaccessible to vehicular traffic, cables installed in heavy gauge steel conduit, the length of span and height above ground being in accordance with Table 11B.

Conduit and cable trunking

18. Rigid conduit supported in accordance with Table 11C.

19. Cable trunking supported in accordance with Table 11D.

20. Conduit embedded in the material of the building.

21. Pliable conduit embedded in the material of the building or in the ground, or supported in accordance with Table 11C.

*Administered in England and Wales by the Secretary of State for Energy and in Scotland by the Secretary of State for Scotland. Obtainable from H.M. Stationery Office.

TABLE 11A

Spacing of supports for cables in accessible positions

Overall diameter of cable‡	Maximum spacing of clips							
	Non-armoured rubber-, p.v.c., or lead-sheathed cables				Armoured cables		Mineral-insulated copper-sheathed or aluminium-sheathed cables	
	Generally		In caravans					
1	Horizontal† 2	Vertical† 3	Horizontal† 4	Vertical† 5	Horizontal† 6	Vertical† 7	Horizontal† 8	Vertical† 9
mm	mm	mm	mm	mm	mm	mm	mm	mm
≤9	250	400			–	–	600	800
>9 ≤15	300	400			350	450	900	1200
>15 ≤20	350	450	150 (for all sizes)	250 (for all sizes)	400	550	1500	2000
>20 ≤40	400	550			450	600	–	–

NOTE – For the spacing of supports for cables of overall diameter exceeding 40mm, and for single-core cables having conductors of cross-sectional area 300mm^2 and larger, the manufacturer's recommendations should be observed.

‡For flat cables taken as the measurement of the major axis.
†The spacings stated for horizontal runs may be applied also to runs at an angle of more than 30° from the vertical. For runs at an angle of 30° or less from the vertical, the vertical spacings are applicable.

TABLE 11B

Maximum lengths of span and minimum heights above ground for overhead wiring between buildings etc.

Type of system 1	Maximum length of span 2	Minimum height of span above ground At road crossings 3	 In positions accessible to vehicular traffic, other than crossings 4	 In positions inaccessible to vehicular traffic† 5
	m	m	m	m
Cables sheathed with p.v.c. or having an oil-resisting and flame-retardant or h.o.f.r. sheath, without intermediate support (Item 15)	3	5.8 for all types)	5.2 for all types)	3.5
Cables sheathed with p.v.c. or having an oil-resisting and flame-retardant or h.o.f.r. sheath, in heavy-gauge steel conduit of diameter not less than 20mm and not jointed in its span (Item 17).	3			3
Bare or p.v.c.-covered overhead lines on insulators without intermediate support (Item 16)	30			3.5
Cables sheathed with p.v.c. or having an oil-resisting and flame-retardant or h.o.f.r. sheath, supported by a catenary wire (Item 13)	No limit			3.5
Aerial cable incorporating a catenary wire (Item 14)	Subject to Item 14			3.5
Bare or p.v.c.-covered overhead lines installed in accordance with the Overhead Line Regulations (Item 16)	No limit			5.2

†This column is not applicable in agricultural premises

NOTE – In some special cases, such as in yacht marinas or where large cranes are present, it will be necessary to increase the minimum height of span above ground given in Table 11B.

TABLE 11C

Spacing of supports for conduits

Nominal size of conduit 1	Maximum distance between supports Rigid Metal Horizontal 2	 Rigid Metal Vertical 3	 Rigid Insulating Horizontal 4	 Rigid Insulating Vertical 5	 Pliable Horizontal 6	 Pliable Vertical 7
mm	m	m	m	m	m	m
Not exceeding 16	0.75	1.0	0.75	1.0	0.3	0.5
Exceeding 16 and not exceeding 25	1.75	2.0	1.5	1.75	0.4	0.6
Exceeding 25 and not exceeding 40	2.0	2.25	1.75	2.0	0.6	0.8
Exceeding 40	2.25	2.5	2.0	2.0	0.8	1.0

NOTES: See p. 179

TABLE 11D

Spacing of supports for cable trunking

Cross-sectional area of trunking	Maximum distance between supports			
	Metal		Insulating	
	Horizontal	Vertical	Horizontal	Vertical
1	2	3	4	5
mm^2	m	m	m	m
Exceeding 300 and not exceeding 700	0.75	1.0	0.5	0.5
Exceeding 700 and not exceeding 1500	1.25	1.5	0.5	0.5
Exceeding 1500 and not exceeding 2500	1.75	2.0	1.25	1.25
Exceeding 2500 and not exceeding 5000	3.0	3.0	1.5	2.0
Exceeding 5000	3.0	3.0	1.75	2.0

NOTES: to Tables 11C and 11D:

1 – The spacings tabulated allow for maximum fill of cables permitted by these Regulations and the thermal limits specified in the relevant British Standards. They assume that the conduit or trunking is not exposed to other mechanical stress.

2 – The above figures do not apply to lighting suspension trunking, or where special strengthening couplers are used. A flexible conduit is not normally required to be supported in its run. Supports should be positioned within 300mm of bends or fittings.

APPENDIX 12

CABLE CAPACITIES OF CONDUIT AND TRUNKING

Introduction

This appendix describes a method which can be used to determine the size of conduit or trunking necessary to accommodate cables of the same size, or differing sizes, and provides a means of compliance with Regulation 529–7.

The method employs a 'unit system', each cable size being allocated a factor. The sum of all factors for the cables intended to be run in the same enclosure is compared against the factors given for conduit or trunking, as appropriate, in order to determine the size of the conduit or trunking necessary to accommodate those cables.

It has been found necessary, for conduit, to distinguish between –

1. straight runs not exceeding 3 metres in length, and

2. straight runs exceeding 3 metres, or runs of any length incorporating bends or sets.

The term 'bend' signifies a British Standard 90° bend, and one double set is equivalent to one bend.

For the case 1, each conduit size is represented by only one factor. For the case 2, each conduit size has a variable factor which is dependent on the length of run and the number of bends or sets. For a particular size of cable the factor allocated to it for case 1 is not the same as for case 2.

For trunking each size of cable has been allocated a factor, as has been each size of trunking.

Because of certain aspects, such as the assessment of reasonable care of pulling-in, acceptable utilisation of the space available and the dimensional tolerances of cables, conduit and trunking, any method of standardizing the cable capacities of such enclosures can only give guidance on the number of cables which can be accommodated. Thus the sizes of conduit or trunking determined by the method given in this appendix are those which can be reasonably expected to accommodate the desired number of cables in a particular run using an acceptable pulling force and with the minimum probability of damage to cable insulation.

Only mechanical considerations have been taken into account in determining the factors given in the following tables. As the number of circuits in a conduit or trunking increases, the current-carrying capacities of the cables must be reduced according to the appropriate grouping factors in Appendix 9. It may therefore be more attractive economically to divide the circuits concerned between two or more enclosures.

This appendix deals with the following four cases:

– Single-core p.v.c.-insulated cables in straight runs of conduit not exceeding 3m in length.

– Single-core p.v.c.-insulated cables in straight runs of conduit exceeding 3m in length, or in runs of any length incorporating bends or sets.

– Single-core p.v.c.-insulated cables in trunking.

– Other sizes and types of cable in trunking.

For other cables and/or conduits not covered by the tables, advice on the number of cables which can be accommodated should be obtained from the manufacturers.

Single-core p.v.c.-insulated cables in straight runs of conduit not exceeding 3m in length.

For each cable it is intended to use, obtain the appropriate factor from Table 12A.

Add all the cable factors so obtained and compare with the conduit factors given in Table 12B.

The conduit size which will satisfactorily accommodate the cables is that size having a factor equal to or exceeding the sum of the cable factors.

TABLE 12A

Cable factors for short straight runs

Type of conductor	Conductor cross-sectional area mm^2	Factor
Solid	1	22
	1.5	27
	2.5	39
Stranded	1.5	31
	2.5	43
	4	58
	6	88
	10	146

TABLE 12B

Conduit factors for short straight runs

Conduit dia mm	Factor
16	290
20	460
25	800
32	1400

Single-core p.v.c.-insulated cables in straight runs of conduit exceeding 3m in length or in runs of any length incorporating bends or sets.

For each cable it is intended to use, obtain the appropriate factor from Table 12C.

Add all the cable factors so obtained and compare with the conduit factors given in Table 12D, taking into account the length of run it is intended to use and the number of bends and sets in that run.

The conduit size which will satisfactorily accommodate the cables is that size having a factor equal to or exceeding the sum of the cable factors.

TABLE 12C

Cable factors for long straight runs, or runs incorporating bends

Type of conductor	Conductor cross-sectional area mm^2	Factor
Solid or stranded	1	16
	1.5	22
	2.5	30
	4	43
	6	58
	10	105

TABLE 12D

Conduit factors for runs incorporating bends

Length of run m	Conduit diameter, mm																			
	16	20	25	32	16	20	25	32	16	20	25	32	16	20	25	32	16	20	25	32
	Straight				One bend				Two bends				Three bends				Four bends			
1					188	303	543	947	177	286	514	900	158	256	463	818	130	213	388	692
1.5	Covered				182	294	528	923	167	270	487	857	143	233	422	750	111	182	333	600
2	by				177	286	514	900	158	256	463	818	130	213	388	692	97	159	292	529
2.5	Tables 12A				171	278	500	878	150	244	442	783	120	196	358	643	86	141	260	474
3	and 12B				167	270	487	857	143	233	422	750	111	182	333	600				
3.5	179	290	521	911	162	263	475	837	136	222	404	720	103	169	311	563				
4	177	286	514	900	158	256	463	818	130	213	388	692	97	159	292	529				
4.5	174	282	507	889	154	250	452	800	125	204	373	667	91	149	275	500				
5	171	278	500	878	150	244	442	783	120	196	358	643	86	141	260	474				
6	167	270	487	857	143	233	422	750	111	182	333	600								
7	162	263	475	837	136	222	404	720	103	169	311	563								
8	158	256	463	818	130	213	388	692	97	159	292	529								
9	154	250	452	800	125	204	373	667	91	149	275	500								
10	150	244	442	783	120	196	358	643	86	141	260	474								

Single-core p.v.c.-insulated cables in trunking

For each cable it is intended to use, obtain the appropriate factor from Table 12E.

Add all the cable factors so obtained and compare with the factors for trunking given in Table 12F.

The size of trunking which will satisfactorily accommodate the cables is that size having a factor equal to or exceeding the sum of the cable factors.

TABLE 12E

Cable factors for trunking

Type of conductor	Conductor cross-sectional area mm²	Factor
Solid	1.5	7.1
	2.5	10.2
Stranded	1.5	8.1
	2.5	11.4
	4	15.2
	6	22.9
	10	36.3

TABLE 12F

Factor for trunking

Dimensions of trunking mm x mm	Factor
50 x 37.5	767
50 x 50	1037
75 x 25	738
75 x 37.5	1146
75 x 50	1555
75 x 75	2371
100 x 25	993
100 x 37.5	1542
100 x 50	2091
100 x 75	3189
100 x 100	4252

For other sizes and types of cable or trunking

For sizes and types of cable and sizes of trunking other than those given in Tables 12E and 12F above, the number of cables installed should be such that the resulting space factor (see Part 2: Definitions) does not exceed 45%.

APPENDIX 13

EXAMPLE OF EARTHING ARRANGEMENTS AND PROTECTIVE CONDUCTORS
(see Chapter 54)

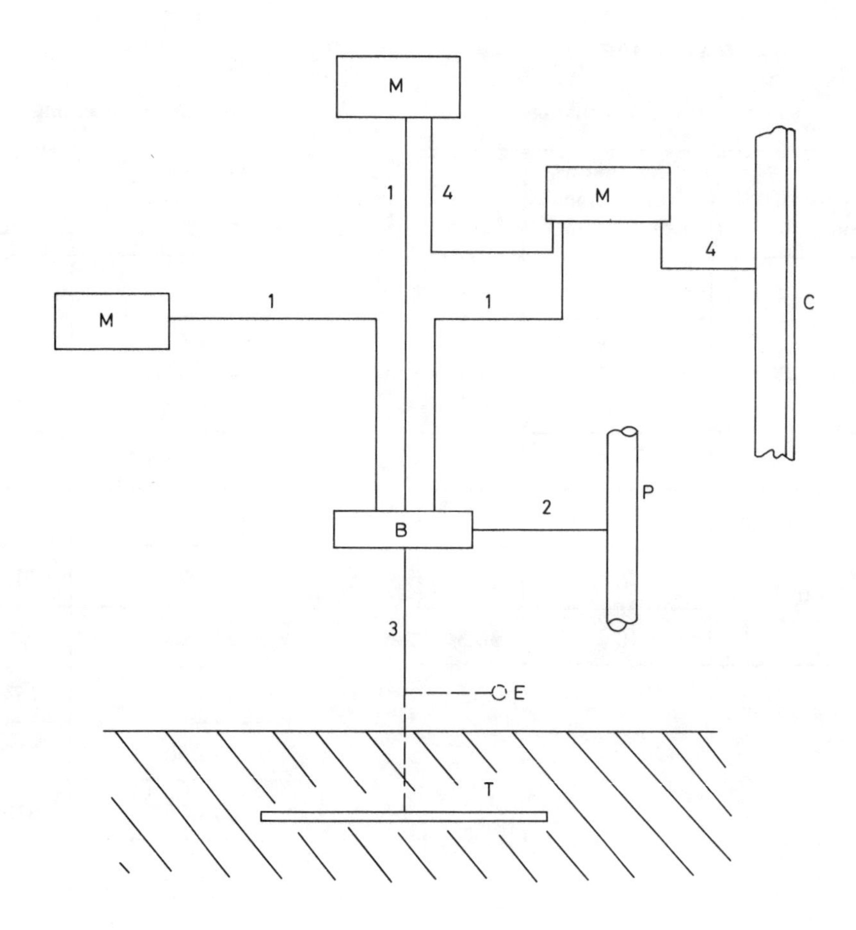

Figure 16

1, 2, 3, 4	=	**protective conductors**
1	=	**circuit protective conductor**
2	=	**main equipotential bonding conductor**
3	=	**earthing conductor**
4	=	**supplementary equipotential bonding conductors (where required)**
B	=	**main earthing terminal**
M	=	**exposed conductive part**
C	=	**extraneous conductive part**
P	=	**main metallic water pipe**
T	=	**earth electrode (TT and IT systems)**
E	=	**other means of earthing (TN systems)**

APPENDIX 14

CHECK LIST FOR INITIAL INSPECTION OF INSTALLATIONS

Visual inspection as required by Regulation 612–1 should include a check of the following items, as relevant to the installation:

– connections of conductors,

– identification of conductors,

– selection of conductors for current-carrying capacity and voltage drop,

– connection of single-pole devices for protection or switching in phase conductors only,

– correct connection of socket outlets and lampholders,

– presence of fire barriers and protection against thermal effects,

– methods of protection against direct contact
(including measurement of distances where appropriate), i.e.:

  – protection by insulation of live parts,
  – protection by barriers or enclosures,
  – protection by obstacles,
  – protection by placing out of reach,
  – protection by non-conducting location.

– presence of appropriate devices for isolation and switching,

– choice and setting of protective and monitoring devices,

– labelling of circuits, fuses, switches, and terminals,

– selection of equipment and protective measures appropriate to external influences,

– presence of danger notices and other warning notices,

– presence of diagrams, instructions and similar information.

APPENDIX 15

STANDARD METHODS OF TESTING

1. General

The methods described in this appendix are suitable for the corresponding testing prescribed in Section 613. They are given as examples, and the use of other methods giving no less effective results is not precluded.

2. Continuity of ring final circuit conductors

Method 1

The continuity of each conductor of the ring circuit, including the protective conductor, is measured between the two ends of the conductor before completion of the ring. The resistance values are noted.

After connection of the two ends of each conductor to complete the ring, the resistance is measured between the corresponding distribution board terminal and the appropriate terminal or contact at the outlet nearest to the midpoint of the ring, using a suitable test lead. The resistance value is noted and the resistance value of the test lead is deducted from this. The resulting value should be approximately one quarter of the corresponding value obtained before completion of the ring.

Method 2

The continuity of each conductor of the ring circuit, including the protective conductor, is measured between the two ends of the conductor before completion of the ring. The resistance values are noted. After completion of the ring, the various conductors of the circuit are all bridged together at the point nearest to the midpoint of the ring. The resistance is then measured between the phase and neutral terminals at the origin of the circuit in the distribution board, when the value obtained should be approximately one half of the value obtained for either the phase or the neutral conductor before completion of the ring.

Where the protective conductor is in the form of a ring, the resistance is then measured between the phase and earth terminals at the origin of the circuit. The value obtained should be the sum of one quarter of the value originally obtained for the phase conductor and one quarter of the value originally obtained for the protective conductor.

3. Continuity of protective conductors and equipotential bonding

The initial tests to be applied to protective conductors are intended to verify that the conductors are electrically sound and correctly connected, before the installation is energised and before any other tests involving these conductors are made. The test is made with a voltage not exceeding 50V a.c. or d.c. and at a current approaching 1.5 times the design current of the circuit under test, except that the current need not exceed 25A. For a.c., the current shall be at the frequency of the supply. If a d.c. test is used, it is to be verified by inspection throughout the length of the protective conductor that no inductor is incorporated. Where the protective conductor is not steel conduit or other steel enclosure the requirement concerning the test current does not apply and a d.c. ohmmeter may be used. It is often more convenient if a hand generator or other portable device is used rather than a transformer fed from the supply, as in this event the live conductors of the various circuits, whilst disconnected from the supply, may be connected for purpose of test to the consumer's earthing terminal and the test can then be made between the phase conductor and the protective conductor at each individual point, such as a socket outlet.

4. Earth electrode resistance

An alternating current of a steady value is passed between the earth electrode T and an auxiliary earth electrode T_1 placed at such a distance from T that the resistance areas of the two electrodes do not overlap. A second auxiliary earth electrode T_2, which may be a metal spike driven into the ground, is then inserted half-way between T and T_1 and the voltage drop between T and T_2 is measured. The resistance of the earth electrodes is then the voltage between T and T_2, divided by the current flowing between T and T_1, provided that there is no overlap of the resistance areas. To check that the resistance of the earth electrodes is a true value, two further readings are taken with the second auxiliary electrode T_2 moved 6m further from and 6m nearer to T respectively. If the three results are substantially in agreement, the mean of the three readings is taken as the resistance of the earth electrode T. If there is no such agreement the tests are repeated with the distance between T and T_1 increased.

The test is made either with current at power frequency, in which case the resistance of the voltmeter used must be high (of the order of 200 ohms per volt), or with alternating current from an earth tester comprising a hand-driven generator, a rectifier (where necessary), and a direct-reading ohmmeter.

If the tests are made at power frequency the source of the current used for the test is isolated from the mains supply (e.g. by a double-wound transformer), and in any event the earth electrode T under test is disconnected from all sources of supply other than that used for testing.

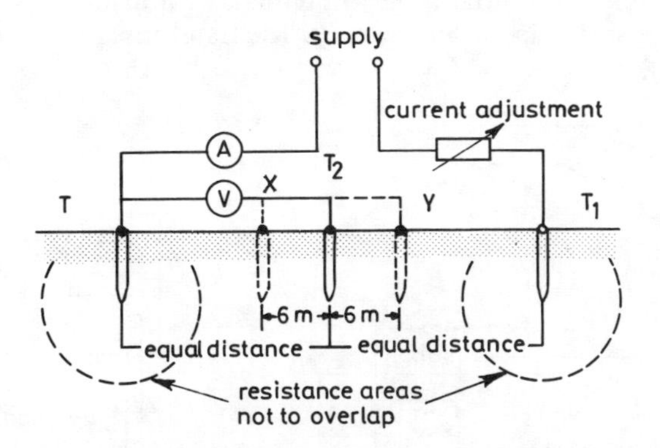

Figure 17 – Measurement of earth electrode resistance (see Item 4 of this appendix)

T – earth electrode under test, disconnected from all other sources of supply

T_1 – auxiliary earth electrode

T_2 – second auxiliary earth electrode

X – alternative position of T_2 for check measurement

Y – further alternative position of T_2 for check measurement.

5. Earth fault loop impedance

The earth fault current loop (phase to earth loop) comprises the following parts, starting at the point of fault:

- the circuit protective conductor, and
- the consumer's earthing terminal and earthing conductor, and
- for TN systems, the metallic return path, or
- for TT and IT systems, the earth return path, and
- the path through the earthed neutral point of the transformer and the transformer winding, and
- the phase conductor from the transformer to the point of fault.

The impedance of the earth fault loop is denoted by Z_s.

These tests may be made using an instrument which measures the current flowing when a known resistance is connected between the phase conductor and consumer's earthing terminal. In using such an instrument, care should be taken that no ill effects can arise if the earthing circuit is defective.

Where a measurement on an a.c. system is made with less than 10 amperes a.c., or with rapidly-reversed d.c., and the protective conductor is wholly or mainly of steel conduit or pipe, the effective value is taken as twice the measured value, less the value measured at the consumer's earthing terminal (Z_E). The value Z_E is that part of the earth fault loop external to the installation. In all other cases the effective value is the measured value. Measurement in systems fed from small transformers may need compensation to include an allowance for the impedance of transformers, balancers and the like.

(i) The measurement of Z_E is made with the main equipotential bonding conductors temporarily disconnected, having verified that the method of test is such that no danger can occur.

(ii) The measurement of Z_s, i.e. at a point distant from the origin of the installation, is made with the main equipotential bonding conductors in place. Alternatively, for an installation having cables of cross-sectional area less than 35 mm^2, Z_s may be obtained by summating arithmetically the measured value of Z_E, the measured value of the resistance of the phase conductor from the origin of the installation or circuit to the most distant socket outlet or other point of utilisation (R_1), and the measured value of the resistance of the circuit protective conductor from the origin of the installation or circuit to that socket outlet or point of utilisation (R_2). The tests of the resistances of the circuit phase and protective conductors may be made using one of the methods described in Item 3 of this appendix. Any instrument used should be suitably calibrated and the test current should be 1.5 times the design current of the circuit under test except that the current need not exceed 25A.

6. Operation of residual current operated and fault-voltage operated protective devices
For this test a voltage not exceeding 50V r.m.s. a.c., obtained from a double wound transformer connected to the mains supply, is applied across the neutral and earth terminals (or neutral and frame terminals for a voltage operated device) and the device shall trip instantaneously. The transformer has a short-time rating not less than 750VA (see Figures 18 and 19).

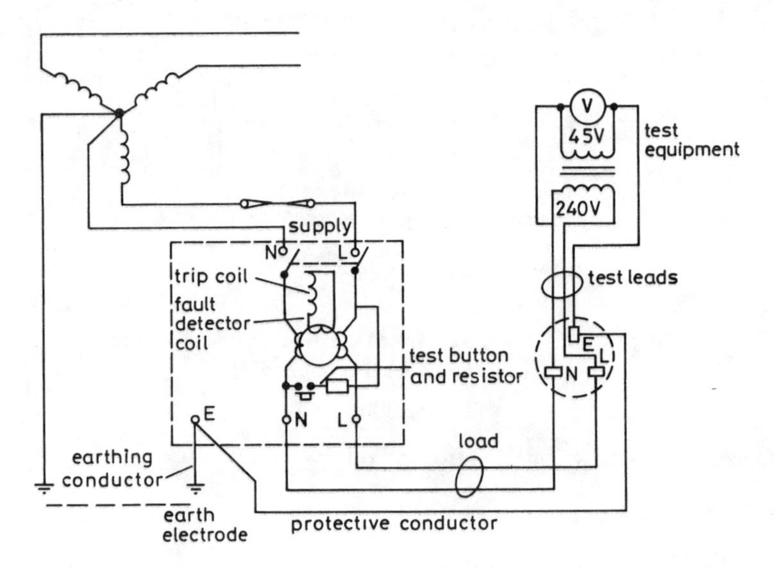

Figure 18 – Method of test for compliance with Regulation 613–16 for a typical residual current operated protective device (see Item 6 of this appendix).

NOTE – A more satisfactory test, but needing more elaborate apparatus, is described in CP 1013.

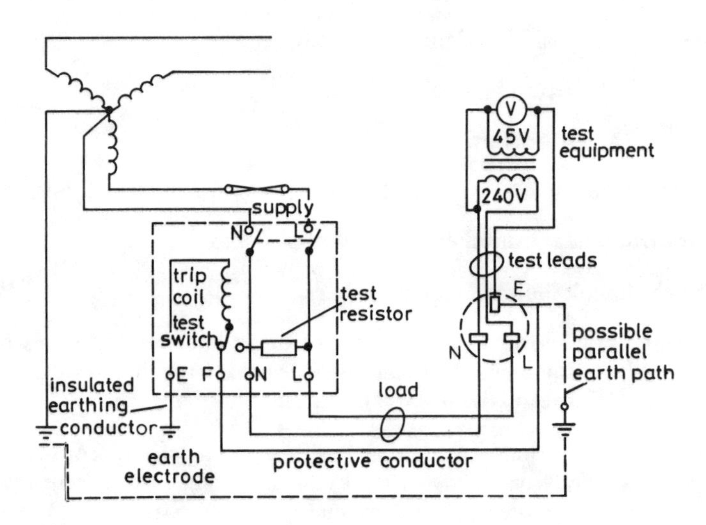

Figure 19 – Method of test for compliance with Regulation 613–16 for a typical fault-voltage operated protective device (see Item 6 of this appendix).

NOTE – A more satisfactory test, but needing more elaborate apparatus, is described in CP 1013.

APPENDIX 16

FORMS OF COMPLETION AND INSPECTION CERTIFICATES

COMPLETION CERTIFICATE

(as prescribed in the IEE Regulations for electrical installations)

Completion Certificate to be given by the contractor or other person responsible for the construction of the installation, or alteration thereto, or by an authorised person acting on his behalf.

I CERTIFY that the electrical installation at:

has been inspected and tested, in accordance with the Regulations for electrical installations, published by the Institution of Electrical Engineers (15th Edition)* and that, to the best of my knowledge and belief, the installation summarised in the drawings/schedule attached/overleaf† complies, at the time of my test, with the Edition of those Regulations current at the date of contract for the work, except as stated overleaf.

I RECOMMEND that this installation be further inspected and tested after an interval of not more than ...... years‡.

Signed .... Date ....

For and on behalf of: ....

....

Address ....

....

NOTE – This Completion Certificate does not cover portable appliances or equipment connected to socket outlets, for which an Inspection Certificate may be obtained.

* See Inspection Certificate attached.

† For simple installations, the particulars of the installation mentioned overleaf are regarded as a sufficient schedule for the purposes of Regulation 514–3.

‡ The space provided in the form for inserting the recommended number of years intervening between inspections should, for installations in general, be filled in by the figure 5 or such lesser figure as is considered appropriate to the individual case. For temporary installations on construction sites, the figure 3 should be inserted and the word 'years' changed to read 'months'. For caravan site installations, the figure should be 1, or such longer period (not exceeding 3 years) as is considered appropriate to the case. For agricultural premises the figure should be 3.

Particulars of the installation covered by this certificate**

New installation. Alteration/Extension to existing installation

Number of lighting points: Number of socket outlets:

Details of fixed current-using equipment:

Applicable ANT Scheme certificate numbers (if any):

Details of departures (if any) from the Regulations:

Comments (if any) on existing installation (where this certificate relates to an alteration or addition):

**Delete or complete items as appropriate to the work.

INSPECTION CERTIFICATE

(as prescribed in the IEE Regulations for electrical installations)

Inspection Certificate to be given by the contractor or other person responsible for carrying out an inspection and test of an installation, or part of an installation, or by an authorised person acting on his behalf.

I CERTIFY that the electrical installation at:

has been inspected and tested, in accordance with the IEE Regulations for electrical installations (15th Edition) and that the results are satisfactory in the respects mentioned below, except as indicated in the comments below.

I RECOMMEND that the installation be further inspected and tested after an interval of not more than . . . . . . years.*

Items inspected or tested:†

Type of earthing arrangements: TN)
TT) (Regulation 312–3)
IT)

Type(s) of protective device:
- overcurrent protective devices,
- residual current device(s),
- fault-voltage operated protective device(s).

Prospective short circuit current at the origin (Regulations 313–1 and 434–2)

Earth fault loop impedance at the origin (Regulation 313–1)

Continuity of ring final circuit conductors (Regulation 613–2)

Continuity of protective conductors and equipotential bonding (Regulation 613–3)

Earth electrode resistance (Regulation 613–4)

Insulation resistance of the fixed installation (Regulations 613–5 to 613–7)

Insulation resistance to earth of each item of equipment tested separately (Regulation 613–8)

Protection against direct contact, by insulation (Regulations 613–9 and 613–10)

Protection against direct contact, by barriers or enclosures (Regulation 613–12)

---

*The space provided in the Certificate for inserting the recommended number of years intervening between inspections should for installations generally, be filled with the figure 5 or such lesser figure as is considered appropriate to the individual case. For temporary installations on construction sites the figure 3 should be inserted and the word 'years' changed to read 'months'. For caravan site installations, the figure should be 1, or such longer period (not exceeding 3 years) as is considered appropriate to the case. For agricultural premises the figure should be 3.

†delete or complete items, as appropriate. Where a failure to comply with the Regulations is indicated further details should be entered, if necessary, overleaf.

Resistance of non-conducting floors and walls, where relied upon for protection against indirect contact (Regulation 613–13)

Polarity, and position of single-pole devices for protection and switching (Regulation 613–14)

Earth fault loop impedance, for operation of devices relied upon for earth fault protection (Regulation 613–15)

Operation of residual current operated/fault-voltage operated device for earth fault protection (Regulation 613–16)

Method of compliance with Regulation 413–3 (see also Regulation 413–4(i))

Protection against indirect contact by measures other than automatic disconnection (Regulations 613–10 and 613–11)

Condition of flexible cables and cords, switches, plugs and socket outlets (Regulation 612–1)

Sizes of live conductors and their methods of installation, in relation to design currents of circuits and to the operating currents of overcurrent protective devices (Regulation 612–1)

Equipment tested includes/does not include portable equipment.

Applicable ANT Scheme certificate numbers (if any):

Comments (if any) and departures from the IEE Regulations:

Signed .... Date ....

For and on behalf of: ....

....

Address: ....

....

....

Printed copies of the inspection and completion certificates cannot be obtained from the Institution. There is, however, no objection to their being reproduced privately in any convenient form, provided the usual acknowledgement of their source is made.

Index

A

Abrasion of cables 523-21, 523-28
AC, types of cable suitable for 521-8
Accessibility–
– cable joints 526-1
– conduit fittings 526-2
– emergency switching 463-4
– equipment 13-16, 513-1
– protective conductor connections 543-16
Accessories–
– non-metallic, fixing screws for 471-26 (iv)
– selection and erection of Sec. 553
Accessory, definition p. 7
Additions to installations 13-19, Chap. 62
Adverse conditions, precautions in 13-17, 13-18
Aerial cables 521-1, Appx. 11
Agricultural and horticultural installations–
– applicability of the Regulations to 11-1
– electric fences 554-40
– overhead wiring Appx. 11 (Table 11B)
– periodic inspection and testing Appx. 16 (footnotes)
– protective measures against shock 471-40, 471-41
– statutory regulations Appx. 2(1)
– wiring systems in livestock locations 523-34
Air conditioning systems, main bonding of 413-2
Aircraft, electrical equipment of, the Regulations not applicable to 11-3 (viii)
Alarm–
– fire (*See Fire alarm*)
– overload 473-3 Note
Alterations to installations 13-19, Chap. 62
Aluminium–
– conductors–
  – as bonding connections 547-1
  – as overhead lines 521-4
  – in contact with brass etc. 523-10
  – less than 16mm^2, prohibited 521-1
  – prohibited for connections to earth electrodes 542-17
– conduit, corrosion of Appx. 10
– sheaths, corrosion of Appx. 10
Ambient temperature–
– cables and conductors 523-1 to 523-6, Appx. 10 (Note 2)
– cables–
  – correction factors Appx. 9
  – determination of Appx. 9 (Preface, 1)
– definition p. 7
Amendments to the Regulations Preface p.vi
Appliance, definition p. 7
Appliances–
– fire hazard from 422-2
– flexible cables for 521-5, 521-6
– heating–
  – and standard final circuits Appx. 5
  – bathrooms 471-39
  – diversity Appx. 4 (Table 4B)
– portable (*See also Portable equipment*)
– portable, connection of 553-9
– stationary, in bathrooms 471-39
– switches mounted on 476-8, 476-18
– switching for 476-17 to 476-20
ANT Scheme 12-6
Application of protective measures for safety Chap. 47
Arcades, discharge lighting in 476-12
Arcing energy, for short circuit protection 434-4
Arcing of switchgear 13-7, 421-2
Area Electricity Boards (*See Supply undertakings*)
Arm's reach, definition p. 7
Arm's reach, zone of 412-10
Assessment of New Techniques Scheme 12-6
Association of Short Circuit Testing Authorities 511-1 Note 4
Automatic disconnection of supply, protection by (*See Earthed equipotential bonding and automatic disconnection*)
Autotransformers, use of 551-1, 551-2, 554-6
Auxiliary supply for residual current device 531-6

B

Bare conductors–
– as overhead lines 412-9
– cables connected to 522-5
– current-carrying capacity 522-4
– electromechanical stresses 522-10
– identification 524-3
– mechanical damage 523-19
– switchboards, on 522-2
– voltage drop 522-8
Bare live parts in SELV circuits 411-10
Bare live parts, placing out of reach 412-10, 412-11
Barrier, definition p. 7
Barriers–
– fire 528-1
– protection by– 412-3 to 412-6
  – application of 471-7, 471-8
  – degrees of protection 412-3, 412-4
  – intermediate, in enclosures 412-6, 413-24
  – openings in 412-3, 471-8
  – provided during erection, testing 613-12
  – removal or opening of 412-6
  – securing of 412-5
– vertical wiring system runs 523-6
Basic insulation, definition p. 7
Bathrooms, protective measures for 471-34 to 471-39
Baths, bonding of 413-7 Note
Batten lampholder, defined as luminaire p. 10

Be–Ca

Bell and call circuits–
– current demand of bell transformer Appx. 4
– Regulations not applicable to parts of 11-3 (iii)
Bends–
– cable ducts or ducting 529-6
– cables 529-3
– conduit systems 529-4, 529-5, Appx. 12
Body resistance–
– conventional 471-2, 471-3 & Note, 471-11 & Note
– in relation to nominal voltage of SELV circuits 471-2, 471-3
– in relation to protection by automatic disconnection 471-11
– livestock 471-40 Note
– reduced 471-2, 471-3 & Note, 471-11 & Note
– very low 471-2, 471-3 & Note, 471-11 & Note
Bonding–
– caravans 471-46
– conductor, definition p. 7
– conductors–
– caravans 471-46
– main 413-2, 547-2, 547-3
– selection and erection of Sec. 547
– supplementary 413-7, 547-4 to 547-7
– connections, warning notice at 514-7
– earth-free local 413-32 to 413-34, 471-20
– equipment in safety separated circuits 413-39
– 'instantaneous' water heaters 554-29
– local supplementary 413-7
– main, to other services and extraneous parts 413-2
– other services at points of contact 525-10
– other services, fundamental requirement 13-11
– single-core metal-sheathed cables 522-7
Boxes–
– joints and terminations in 527-5
– non-metallic, for suspension of luminaires 523-5
– segregation of circuits in 525-7
Breaking capacity–
– fundamental requirement 13-7
– overload protective devices 432-2, 432-3
– short-circuit protective devices 432-2, 432-4, 434-4 434-5
British Approvals Service for Electric Cables 511-1 Note 4
British Electrotechnical Approvals Board 511-1 Note 4
British Electrotechnical Committee Preface p. vi
British Standards– Preface p. vi
– Codes of Practice Preface p. vi
– compliance with 511-1, 612-1
– Institution Preface p. vi, Appx. 1
– Institution certification schemes 511-1 Note 4
– mentioned in the Regulations, list Appx. 1
Building materials, fire hazard from equipment 422-1, 422-2
Building site (*See Construction site*)
Building Standards (Scotland) Regulations Appx. 2
Bunched, definition p. 7
Busbars and busbar connections–
– cables connected to busbars 522-5
– colour identification of 524-3
– current-carrying capacity 522-2
– selection of 521-3
– trunking systems, as protective conductors 543-7
Burns, protection against 423-1

C

Cable–
– capacities of conduit and trunking 529-7, Appx. 12
– channels, temperatures in 523-6
– channels, waterproofing 523-15
– coupler, definition p. 7
– couplers–
– for joints in flexibles 527-9
– reduced low voltage circuits 471-33
– selection and connection of 553-10 to 553-12
– covers 523-23
– ducting–
– bends 529-6
– definition p. 7
– fire barriers in 528-1
– joints in 527-12, 543-16
– protective conductors formed by 543-9, 543-10, 543-13
– selection of 521-12
– supports 529-2
– terminations 527-5
– vertical, temperatures in 523-6
– duct, definition p. 8
– ducts–
– bends 529-6
– cast in situ 521-14, 523-22
– concrete 521-14, 523-22, Appx. 10
– fire barriers in 528-1
– joints in 527-12
– underground 523-24
– vertical, temperatures in 523-6
– waterproofing 523-8, 523-13, 523-15
– enclosures–
– a.c. circuits in steel 521-8
– cable capacities of 529-7, Appx. 12
– trunking (*See Trunking systems*)
Cables–
– abrasion 523-21, 523-28
– a.c., types for 521-8
– aluminium-conductor–
– as bonding connections 547-1
– less than 16mm², prohibited 521-1
– prohibited for connections to earth electrodes 542-17
– aluminium-sheathed, corrosion of Appx. 10
– ambient temperature 523-1 to 523-6, Appx. 9

Cables – *continued*
– armour as protective conductor 543-8
– attack by vermin 523-34
– bends 529-3 to 529-6
– buried direct in ground 523-23
– colour identification Sec. 524
– concealed under floors or above ceilings 523-20
– connected to bare conductors or busbars 522-5
– core identification Sec. 524
– corrosive or polluting substances in contact with 523-17, 523-18, Appx. 10
– current-carrying capacity 522-1 to 522-7, Appx. 9
– damage by fauna 523-34
– electromagnetic effects 521-8
– electromechanical stresses 522-10
– emergency lighting, segregation 525-1 to 525-9
– enclosures, in onerous dust conditions 523-16
– environmental conditions Sec. 523
– extra-low voltage 521-7
– fire-alarm, segregation 525-1 to 525-9
– fire barriers 528-1
– fixings (*See Cables, Supports*)
– flexible (including flexible cords)–
  – appliances, for 521-5, 521-6 & Notes
  – armoured 521-5, 523-27
  – caravans, in 554-2
  – colour identification 524-4, 524-5
  – cords, applications of Appx. 10
  – current-carrying capacity Appx. 9
  – fixed equipment, connections to 523-31
  – fixed wiring, use as 523-29
  – glass-fibre-insulated cords, limitations 523-28
  – identification of cores 524-4, 524-5
  – joints 527-9
  – length of 523-30, 523-31, 553-9 Note
  – luminaires, for 521-5, 521-6 & Notes
  – luminaires, supporting 523-32
  – mass supported by cords 523-32
  – mechanical damage 523-27
  – portable equipment to be connected by 523-30
  – safety separated circuits 413-37, 413-39
  – screened 521-5
  – segregation of circuits in 525-9
  – selection of types, LV 521-5, 521-6, Appx. 10
  – supporting luminaires 523-32
  – unkinkable cords 523-27
– floor-warming 554-31 to 554-34
– glands 527-8
– grouping Appx. 9 (Preface & Table 9D)
– heated floors etc., in 523-2
– heating 554-31 to 554-34
– HV discharge lighting (*See High voltage*)
– identification of cores Sec. 524
– joints, accessibility 526-1
– joints, general Sec. 527
– lift shafts 525-12
– livestock situations 523-34
– low temperatures, in 523-4, Appx. 10 Note

Cables – *continued*
– mechanical stresses 523-19 to 523-33, Appx. 10
– metal coverings as protective conductors 543-8
– metal-sheathed, single-core, bonding 522-7
– metal sheaths as earth electrodes 542-10, 542-15
– methods of installation 521-13, 521-14, Appces. 9 & 10
– methods of installation, typical Appx. 9 (Table 9A)
– mineral-insulated–
  – discharge lighting circuits 521-1 Note 3
  – emergency lighting systems 525-6
  – fire alarm systems 525-6
  – sealing 523-12
  – terminations 527-7
– non-flexible, LV, selection of types 521-1, 521-2
– non-metal-enclosed, as protection against direct and indirect contact 471-18
– non-sheathed, prohibited in concrete ducts 523-22
– non-sheathed, to be enclosed 523-13, 523-22
– numbers in enclosures 529-7
– operational conditions Sec. 522
– outdoors on walls 523-25
– overhead between buildings etc. 523-26, Appx. 11 (13, 14, 15, 17)
– paper-insulated, non-draining 521-2
– paper-insulated, sealing 523-11
– paralleled 433-3, 434-7, 522-3
– reduced neutral 522-9
– road-warming 554-31 to 554-34
– segregation of circuits 525-1 to 525-9
– segregation from other services 525-10 to 525-12
– selection and erection Chap. 52, Appx. 10
– single-core metal-sheathed, bonding 522-7
– soil-warming 554-31 to 554-34
– soldering of 523-18
– space factors 529-7
– sunlight, exposed to 523-35
– supports, general 529-1, 529-2, Appx. 11
– supports, HV discharge lighting (*See High voltage*)
– telecommunication, segregation 525-1 to 525-9
– temperatures for (*See Temperature*)
– terminations Sec. 527
– thermal insulation, in 522-6
– trenches, correction factors Appx. 9 (Table 9C)
– underground 523-23, 523-24, Appx. 10
– voltage drop 522-8, Appx. 9
– water or moisture in contact with 523-7 to 523-15

Capacitive electrical energy, discharge of 461-4
Capacitive equipment, switches for 512-1 Note 1, 512-2 Note

Capacitors, HV discharge lighting 554-5

Ca–Cl

Caravan, definition p. 7
Caravan site, definition p. 7
Caravan site installations–
– applicability of the Regulations to 11-1
– periodic inspection and testing Appx. 16 (footnotes)
– plugs and socket outlets for 553-6
– protective measures for 471-42 to 471-46
– site licensing requirements Appx. 2(8)
Caravan Sites and Control of Development Act Appx. 2(8)
Caravans–
– bonding 471-46
– cable supports Appx. 11 (9, 10, Table 11A)
– connectors 553-13
– couplers 553-13
– inlets 553-13
– luminaires 554-2
– multi-unit (*See definition p. 7*)
– protective measures for 471-42 to 471-46
– warning notice in 514-6
Cartridge fuse link, definition p. 7
Catenary wires 521-1, Appx. 11 (13, 14)
Ceiling roses–
– exemption from requirements for enclosures 412-6
– multiple pendants 553-20
– voltage limit 553-19
Ceilings, thermally insulated, cables above 522-6
Ceilings, cables above, mechanical damage 523-20
CENELEC Preface p. vi
Central heating risers, main bonding to 413-2
Certificates–
– completion 614-1, 622-1, Appx. 16
– forms of Appx. 16
– inspection 614-1 Notes, 631-1, Appx. 16
– provision of 614-1
– recommendation on periodic inspection and testing Appx. 16 (footnotes)
Certification Secs. 614, 622
Certification schemes 511-1 Note 4
Characteristics, general, of installation 300-1
Charts (*See Diagrams*)
Cinematograph installations Appx. 2 (1, iv)
Circuit arrangements–
– division of installation 314-1
– standard Appx. 5
– standard, diversity not applicable to Appx. 4
Circuit breaker, definition p. 8
Circuit breakers–
– adjustable, precautions against interference with setting 533-5
– capacitive equipment, controlling 512-1 Note 1, 512-2 Note
– electrode water heaters, for 554-21, 554-23, 554-26
– emergency switching by, to fail safe 537-14

Circuit breakers – *continued*
– inductive equipment, controlling 512-1 Note 1, 512-2 Note
– miniature, as protection against shock 413-5, 413-9, 413-12
– nominal current to be indicated 533-1
– overcurrent settings, precautions against interference with 533-5
– overload protection by 432-2, 432-3
– position of, fundamental requirements 13-12, 13-13
– short-circuit protection by 432-2, 432-4, 434-5 Note
– voltage rating for fault conditions 530-3
Circuit, definition p. 7
Circuit disconnecting times (*See Disconnecting times*)
Circuit impedances, for automatic disconnection 413-3, 413-5
Circuit protective conductor, definition p. 8
Circuit protective conductor, of ring circuit 543-6
Circuits–
– 'Class II', protective conductors in 413-25, 471-17
– division of installation into 314-1
– final (*See Final circuits*)
– isolation Sec. 461
– number of 314-1
– protective conductor impedances for Appx. 7
– ring (*See Final circuits, ring*)
– safety separated, arrangement of 413-37
– SELV, arrangement of 411-4
– separation of 314-1, 411-4, 546-3
– switching of 476-16
Class I equipment, definition p. 8
Class II equipment, definition p. 8
Class II equipment, treated as Class I where connected to protective conductor 471-17
Class II equipment or equivalent insulation, protection by–
– application of 471-16 to 471-18
– application to complete installation or circuit 471-18
– conductive parts, connection to protective conductor prohibited 413-25
– enclosures of equipment 413-20 to 413-26
– insulating screws not to be relied upon 413-23
– lids or doors in enclosures 413-24
– preferred, for agricultural installations 471-40
– protective conductors in relation to 413-25, 471-17
– reinforced insulation during erection 413-18 (iii)
– supplementary insulation during erection 413-18 (ii)
– testing 413-22, 613-10
– type-tested equipment 413-18 (i)

Cl–Co

Clearances and creepages, discharge lighting (*See High voltage*)
Clocks, current demand of Appx. 4
Clocks, plugs and socket outlets for 553-5
CNE conductor (*See PEN conductor*)
Coal mine installations Appx. 2 (1, v)
Codes of Practice Preface p. vi, Appx. 1
Colour–
– emergency switching devices 537-15
– fireman's switch 537-17
– identification of cables and conductors Sec. 524
– identification of conduits 524-2
Committee, Wiring Regulations, constitution p. v
Compatibility of equipment 331-1, 512-5
Completion certificate–
– form of Appx. 16
– provision of 614-1, 622-1
– recommendation on periodic inspection and testing Appx. 16 (footnote)
Concrete ducts, cables in 521-14, 523-22, Appx. 10
Concrete, steel reinforcement (*See Steel*)
Conductive parts (*See also Exposed conductive parts, Extraneous conductive parts*)
Conductive parts in 'Class II' enclosures 413-25
Conductors–
– aluminium (*See Aluminium conductors*)
– ambient temperature 523-1 to 523-6
– bare, cables connected to 522-5
– bare, current-carrying capacity 522-4
– bare, on switchboards 522-2
– bonding (*See Bonding conductors*)
– copperclad (*See Aluminium conductors*)
– corrosive or polluting substances, exposure to 523-17, 523-18
– colour identification of 524-3
– current-carrying capacity 13-4, 522-1 to 522-7, Appx. 9
– electromechanical stresses 522-10
– environmental conditions Sec. 523
– erection of, fundamental requirement 13-5
– floor-warming 554-31 to 554-34
– HV discharge lighting (*See High voltage*)
– identification Sec. 524
– insulation of, fundamental requirement 13-5
– joints Sec. 527
– mechanical stresses 523-19 to 523-33
– methods of installation 521-13
– neutral, cross-sectional area 522-9
– operational conditions Sec. 522
– overhead (*See Overhead lines, Overhead wiring*)
– paralleled 433-3, 434-7, 522-3
– protective (*See Protective conductors*)
– road-warming 554-31 to 554-34
– selection and erection Chap. 52
– soil-warming 554-31 to 554-34
– soldering of 523-18
– supports, general 529-1, 529-2, Appx. 11
– supports, HV discharge lighting (*See High voltage*)
– switchboard 522-1, 522-2

Conductors – *continued*
– terminations Sec. 527
– voltage drop 522-8
– water or moisture, exposed to 523-7 to 523-15
Conduit and conduit systems–
– aluminium, corrosion of Appx. 10
– bends 529-5, Appx. 12
– boxes 527-10, 527-11, 543-10
– cable capacities of 529-7, Appx. 12
– colour identification of 524-2
– drainage of 523-14
– elbows 529-4
– erection to be completed before drawing in 521-10
– fixing (*See Supports*)
– flexible, prohibited as protective conductor 543-11
– joints 527-10 to 527-12, 543-16, 543-19
– metallic–
  – pin-grip and plain slip fittings prohibited 543-19
  – protective conductor formed by 543-9 to 543-11, 543-13
  – socket outlets in 543-10
– non-metallic, ambient temperature for 523-5
– non-metallic, boxes suspending luminaires 523-5
– outdoors, on walls 523-25
– overhead between buildings etc. 523-26, Appx. 11 (17)
– passing through floors and walls 528-1
– prefabricated 521-11
– protective conductor formed by 543-9 to 543-11, 543-13
– selection of types 521-9
– short lengths for mechanical protection of cables 471-26 (v)
– space factors in 529-7, Appx. 12
– supports 529-2, Appx. 11
– tees 529-4
– terminations 527-5, 527-10
– underground, cables for 523-24
– water or moisture in contact with 523-8
Confined conductive location–
– definition p. 8
– earth fault loop impedance in 471-11 & Note
– nominal voltage of SELV circuits in 471-2, 471-3 & Note
Connection, electrical, fundamental requirements 13-6
Connector, definition p. 8
Connectors, caravan 553-13
Connectors, clock 553-5
Construction of equipment, applicability of the Regulations to 11-4
Construction (General Provisions) Regulations Appx. 2
Construction site installations–
– applicability of the Regulations to 11-1
– cable couplers 553-12

Construction site installations – *continued*

– cable supports Appx. 11 (12)
– periodic inspection and testing Appx. 16 (footnotes)
– plugs and socket outlets 553-7
– statutory regulations Appx. 2
Contactors, emergency switching by, to fail safe 537-14
Continuity tests 613-2, 613-3, Appx. 15 (2, 3)
Contracts, citation of the Regulations in 12-2
Control panels, connections to, omission of short-circuit protection 473-8
Controlgear, labels and indicators for 514-1
Controls in bathrooms 471-39
Cooking appliances–
– current demand Appx. 4
– diversity Appx. 4 (Table 4B)
– household, standard final circuits Appx. 5
– switching 476-20
Copperclad aluminium conductors (*See Aluminium conductors*)

Cord-operated switches in bathrooms 471-39
Correction factors–
– ambient temperature for cables Appx. 9
– cables in trenches Appx. 9 (Table 9C)
– grouping of cables Appx. 9 (Table 9B)

Corrosion–
– cables, general 523-17, Appx. 10
– earth electrodes 542-12, 542-13
– earthing conductors 542-18
– heating cables 554-32
– products of overheated p.v.c 523-3 Note
Corrosive atmospheres, equipment exposed to 13-17
Corrosive substances, wiring systems exposed to 523-17
Coupler, cable (*See Cable coupler*)
Coupler, caravan 553-13

Creepages and clearances, in HV discharge lighting circuits (*See High voltage*)
Creosote, in contact with non-metallic wiring systems 523-17
Current–
– nature of, assessment 313-1
– nominal, of fuses and circuit breakers 533-1
– suitability of equipment for 512-2
Current-carrying capacity–
– cables and conductors 13-4, 522-1 to 522-7, Appx. 9
– definition p. 8
– parallel conductors 433-3, 522-3

Current demand (*See Maximum demand*)
Current transformers, omission of overload protection 473-3
Current-using equipment–
– definition p. 8
– power demand 13-3
– selection and erection of Sec. 554

D

Danger, definition p. 8
Danger notices (*See Warning notices*)
Data transmission circuits, applicability of the Regulations to 11-3 (iii)
Date of validity of the Regulations 12-9
D.C. feedback, assessment for compatibility 331-1
D.C. plugs and socket outlets, prohibited for switching 537-18
Defects or omissions, on test and inspection 614-1, 622-1
Defined conditions, for cable current-carrying capacities Appx. 9 (Table 9A)
Definitions Part 2
Demand, maximum, assessment of 311-1, Appx. 4
Demand, maximum, suitability of supply for 313-1
Departures from the Regulations 12-3, 12-5, Appx. 2
Design current, definition p. 8
Diagrams, provision of 514-3
Diagrams, availability for inspection and testing 611-2
Dielectric, flammable liquid 422-5
Direct and indirect contact, protection against both– Secs. 411, 471
– application of 410-1, 471-2 to 471-5
– cables as 471-18
– in areas accessible only to skilled or instructed persons 471-22 to 471-25
Direct contact, definition p. 8
Direct contact, protection against– Secs. 412, 471
– application of 471-6 to 471-10
– in areas accessible only to skilled or instructed persons 471-22 to 471-25
– supplementary, by residual current device 471-14 Note
– testing of 613-9, 613-12
Discharge lighting–
– applicability of the Regulations to 11-2
– circuits, mineral-insulated cables in 521-1 Note 3
– circuits, reduced neutral prohibited 522-9
– current demand Appx. 4
– fireman's switch 476-12 to 476-13, 537-17
– high voltage (*See High voltage*)
– inductors 551-4
– isolation 476-6
– switches 537-19
– transformers 551-4, 551-5
Discharge of energy, for isolation 461-4
Discharge of energy, protection by limitation of 411-16, 471-5
Disconnecting times–
– complying with temperature limits for protective conductors Appx. 8
– equipment in bathrooms 471-36
– reduced low voltage circuits 471-32

Disconnecting times – *continued*
– shock protection, fixed equipment outside main equipotential zone 471-12
– shock protection, general 413-4
Discrimination, fault-voltage operated devices 544-2
Discrimination, overcurrent protective devices 533-6
Disputes with supply undertakings Appx. 2
Distribution boards, backless, erection of 422-3
Diversity, application of Appx. 4
Diversity, assessment of 311-2
Doors in 'Class II' enclosures 413-24
Double insulation, definition p. 8
Double insulation, of equipment 413-18 to 413-26
Drainage of conduit systems 523-14
Duct, definition p. 8
Duct (*See also Cable ducts*)
Ducting (*See Cable ducting*)
Dust conditions, onerous 523-16
Dynamic stresses, for short-circuit protective devices 434-4

E

Earth–
– connections to 331-1, Sec. 542
– definition p. 8
– protection of persons in contact with general mass of 471-12
Earth electrode–
– caravan sites 471-43
– definition p. 8
– independent, for fault-voltage operated device 544-2
– IT systems 413-14
– resistance–
– allowance for soil drying and freezing 542-11
– allowance for corrosion 542-13
– area, definition p. 11
– areas, not to overlap 544-2
– definition p. 8
– for protection against shock 413-6
– test of 613-4, Appx. 15(4)
Earth electrodes–
– electric fences 554-38
– electrically independent, definition p. 9
– selection of 542-10 to 542-15
Earth fault current–
– adequacy of earthing arrangements for 542-7
– fundamental requirements 13-8, 13-10
– method of calculating Appx. 8
– path, responsibility of installer for Appx. 3
Earth fault loop impedance–
– agricultural installations 471-40
– body resistance, in relation to 471-11 & Note
– definition p. 9
– equipment used outside main equipotential zone 471-12

Earth fault loop impedance – *continued*
– external to installation, assessment 313-1
– for automatic disconnection for protection against indirect contact 413-5, 413-6
– limitation, for calculating sizes of protective conductors Appx. 8
– measurement 613-15
– reduced low voltage circuits 471-32
– testing 613-15, Appx. 15(5)
Earth-free local equipotential bonding, protection by– 413-32 to 413-34, 471-20
– application of 471-20
– avoidance of contact with Earth 413-33
– bonding of parts 413-32
– limited to special situations 471-20
– precautions at entrance to location 413-34
Earth-free location (*See Non-conducting location*)
Earth leakage current–
– assessment for compatibility 331-1
– definition p. 9
– fundamental requirements 13-10
– HV discharge lighting circuits 554-4
– HV electrode water heaters and boilers 554-23
– residual current devices and 531-5
– suitability of earthing arrangements for 542-7
Earth loop impedance (*See Earth fault loop impedance*)
Earth monitoring 543-18
Earthed concentric wiring, definition p. 9
Earthed concentric wiring (*See also PEN conductors*)
Earthed equipotential bonding and automatic disconnection, protection by– 413-2 to 413-17, 471-11 to 471-15
– agricultural installations 471-40
– application of 471-11 to 471-15
– basic requirements 413-3
– body resistance in relation to 471-11
– caravan installations 471-42, 471-45
– coordination of characteristics 413-3
– devices for, types 413-9, 413-12, 413-15
– disconnecting times 413-4, 471-12, 471-32, 471-36
– earth electrode resistance, for fault-voltage operated device 413-6
– earth fault loop impedances 413-5
– equipment used outside main equipotential zone 471-12
– exposed conductive parts, connection of–
– IT systems 413-14
– TN systems 413-8

Earthed equipotential bonding and automatic disconnection, protection by – *continued*
– TT systems 413-10, 413-11
– fault-voltage operated device for 471-15
– local supplementary bonding 413-7
– main equipotential bonding 413-2
– reduced system voltages and 471-27 to 471-33
– residual current device for 471-14
– socket outlet circuits in TT systems 471-13
Earthing–
– arrangements–
– determination of type of 312-3
– example of Appx. 13
– responsibility of installer for Appx 3
– selection and erection Chap. 54
– caravan sites 471-43
– conductor, definition p. 9
– conductors, electrode boilers 554-22
– conductors, selection and erection 542-16 to 542-18, 544-4
– connections, warning notices at 514-7
– exposed conductive parts–
– IT systems 413-14
– TN systems 413-8
– TT systems 413-10, 413-11
– functional (*See Functional earthing*)
– lightning protection systems 541-2 Note
– neutral, in reduced LV circuits 471-30
– neutral of supply 471-30, 546-1, 554-24, 554-25, Appx. 3
– prohibited, in earth-free local bonded location 413-33
– prohibited, in non-conducting location 413-28
– resistance–
– provision for measurement of 542-20
– selection of value of 542-7
– variations in 531-5, 531-9, 542-11, 542-13
– responsibility of installer for Appx. 3
– supply system, explanatory notes on types of Appx. 3
– terminal, main–
– connection to Earth 542-1 to 542-9
– definition p. 10
– selection and erection 542-19, 542-20
Editions, list of p. iv
Effects of the Regulations Chap. 12
Electric braking, with emergency stopping 537-13
Electric shock–
– definition p. 9
– in case of a fault (*See Indirect contact*)
– in normal service (*See Direct contact*)
– protection against Chap. 41, Sec. 471
– current (*See Shock current*)
Electric traction equipment, the Regulations not applicable to 11-3 (iv)
Electrical engineer (*See Engineer*)
Electrical equipment, definition p. 9
Electrical equipment (*See also Equipment*)
Electrical Equipment (Safety) Regulations–
– on types of flexible cord 521-6 Note 1
– relationship of the Regulations to 12-3, Appx. 2
Electrical installation, definition p. 9
Electrical separation, protection by– 413-35 to 413-39, 471-21
– application of 471-21
– bonding of equipment 413-39
– exposed metalwork of circuit 413-38, 413-39
– flexible cables for 413-39, 413-37
– for one item of equipment 413-35 to 413-38
– for several items of equipment 413-35 to 413-37, 413-39, 471-21
– protective conductors for 413-39
– separation of circuit 413-37
– socket outlets for 413-39
– supplies for 413-36
– testing of 613-11
– voltage limitation 413-36
Electrically independent earth electrodes–
– definition p. 9
– for fault-voltage operated device 544-2
Electricity (Factories Act) Special Regulations–
– applicable to certain caravans Appx. 2
– applicable to construction sites Appx. 2
– areas accessible to skilled persons 471-24
– passageways and working platforms 471-23
– relationship of the Regulations to 12-3, Appx.2
– requirements for enclosures 412-3 Note
Electricity Supply Regulations–
– cable types approved for 521-1(ix) & Note 2
– connection of neutral with Earth 13-13 Note, 554-24 Note, Appx. 3
– earth leakage current 13-9 Note
– relationship of the Regulations to 12-3, Appx. 2
Electrode boilers (or electrode water heaters)–
– applicability of the Regulations to 11-2
– definition p. 9
– selection and erection of 554-20 to 554-26
Electrodynamic effects on protective conductors 543-15
Electrodynamic effects (*See also Electromechanical, Electromagnetic*)
Electrolysis, precautions against–
– earthing arrangements 542-8
– wiring system metalwork Appx. 10
Electromagnetic effects in cables 521-8
Electromagnetic equipment, switching for mechanical maintenance 476-7

Electromechanical stresses–
– conductors and cables 522-10
– earthing arrangements 542-7
– protective conductors 543-15
Electronic devices– (*See also Semiconductors*)
– as safety source for SELV circuit 411-3
– testing of circuits incorporating 613-7
Emergency lighting systems–
– applicability of the Regulations to 11-3 (iii)
– segregation from LV circuits 525-1 to 525-9
– supplies for 313-2
Emergency stopping 463-5, 537-13
Emergency switching–
– application of 476-9 to 476-13
– definition p. 9
– provision of Sec. 463
– selection of devices for 537-12 to 537-17
Enclosure, definition p. 9
Enclosures–
– effect on accessibility 513-1
– joints and terminations, for 527-4
– protective conductors formed by 543-7, 543-9
– provided during erection, testing of 613-10, 613-12
Enclosures, protection by–
– Class II or equivalent 413-18 to 413-26
– degrees of protection 412-3, 412-4
– fire-resistant, for fixed equipment 422-2
– lids or doors in 413-24
– openings in 412-3
– provided during erection, testing of 613-10, 613-12
– removal or opening of 412-6
– securing of 412-5, 413-19
Energy, limitation of discharge of 411-16
Engineer, suitably qualified electrical 12-2, 314-1, 471-19 to 471-21, 521-13, Appx. 4
Environmental conditions, of cables and conductors Sec. 523
Environmental conditions (*See also External influences*)
Equipment–
– accessibility 13-6, 513-1
– agricultural installations 471-40
– applicability of the Regulations to 11-4
– capacitive, switches for 512-1 Note 1, 512-2 Note
– Class II 413-18 to 413-26, 471-16 to 471-18
– compatibility of 512-5
– construction of 13-2
– containing flammable dielectric liquid 422-5
– current-using, definition p. 8
– current-using, requirements 13-3, Sec. 554
– definition p. 9
– electrically heated, mechanical maintenance switching for 476-7
– erection of Part 5
– fixed–
  – definition p. 9
  – disconnecting times for 413-4

Equipment – *continued*
– earth fault loop impedances for 413-5
– heat dissipation 422-1
– fundamental requirements 13-2
– guarding, against burns 423-1
– having exposed live parts 471-23
– heating, mechanical maintenance switching 476-7
– incorporating motors 552-3
– inductive, switches for 512-1 Note 1, 512-2 Note
– isolation and switching of Chap. 46
– maintenance of 13-2, 341-1
– mechanical maintenance switching for Sec. 462, 476-7 and 476-8
– mobile, in non-conducting location 413-30
– oil-filled 422-5
– operational conditions Sec. 512
– outdoors, circuits supplying 471-12
– portable–
  – connection of 553-9
  – definition p. 11
  – in non-conducting location 413-30
  – prohibited in bathrooms 471-34
– selection of Part 5
– stationary, current demand of Appx. 4
– stationary, definition p. 12
– suitability for maximum power demand 13-3
– surface temperature of 422-2, 423-1
– testing of 13-2, Part 6
– type-tested 413-18 (i)
– used outside main equipotential zone 471-12
– ventilation of 422-2
Equipotential bonding–
– caravans 471-46
– conductors–
  – main 413-2, 547-2, 547-3
  – selection and erection of Sec. 547
  – supplementary 413-7, 547-4 to 547-7
– definition p. 9
– earth-free local 413-32 to 413-34, 471-20
– local supplementary, bathrooms 471-35
– local supplementary, general 413-7
– main 413-2
– safety separated circuits, in 413-39
– testing of 613-3
– zone, equipment used outside 471-12
ERA Technology Ltd. Appx. 9
Erection of equipment Part 5
Established materials and methods, use of 12-4
European Committee for Electrotechnical Standardization (CENELEC) Preface p. vi
Exciter circuits of machines, omission of overload protection 473-3
Exhibition buildings, discharge lighting in 476-12

Explosion hazard, the Regulations not applicable to 11-3 (ii)
Explosive atmospheres, installations in–
– applicability of the Regulations to 11-3 (ii)
– equipment in, fundamental requirement 13-18
– statutory regulations for Appx. 2 (6)
Exposed conductive part, definition p. 9
Exposed conductive parts–
– bathrooms 471-35, 471-36
– connection, for earthing and automatic disconnection–
  – IT systems 413-14
  – TN systems 413-8
  – TT systems 413-10, 413-11
– earth-free local bonded location 413-32
– exemptions from protection against indirect contact 471-26 (iii)
– FELV systems 411-14
– local bonding of 413-7
– protective conductors formed by 543-12
– SELV circuits 411-5
External influence, definition p. 9
External influences– (*See also Environmental conditions*)
– classification of Appx. 6
– earthing arrangements to be suitable for 542-7
– equipment to be suitable for 512-6
– explanatory note on Chap. 32
Extra-low voltage– (*See also FELV and SELV*)
– cables for 521-7
– definition p. 12
– range 11-2
Extraneous conductive part, definition p. 9
Extraneous conductive parts–
– bathrooms 471-35, 471-36
– bonding of 413-2, 413-7
– caravans 471-46
– connection to Earth–
  – IT systems 413-14
  – TT systems 413-11
– propagating potential outside non-conducting location 413-31
– protective conductors formed by 543-14, 547-7, 613-3
– SELV crcuits 411-5

F

Factory-built assembly (of LV switchgear)–
– definition p. 9
– enclosure used as protective conductor 543-7
– having total insulation 413-18 (i)
Factory Regulations (*See Electricity (Factories Act) Special Regulations)*
Fairgrounds, applicability of Regulations to 11-5
Farms (*See Agricultural installations*)
Fault-voltage operated protective devices–
– application in TT and IT systems, for shock protection 471-15
– earth electrode resistance for 413-6
– earth fault loop impedance for 413-6
– earthing, and protective conductors for Sec. 544
– selection of 531-9
– shock protection by 413-6, 413-11, 413-12, 413-15
– testing 613-16, Appx. 15 (6)
Fauna, damage to cables by 523-34
Feedback of d.c., assessment for compatibility 331-1
FELV systems–
– application of 471-4
– exposed conductive parts of 411-14
– general 411-11
– insulation and enclosure of 411-12, 411-13
– plugs and socket outlets of 411-15
Fence controllers, as example of limitation of discharge of energy 471-5
Fence controllers, selection and erection of 554-35 to 554-40
Fences, electric, erection of 554-40
Final circuit, definition p. 9
Final circuits–
– arrangement of, general 314-3
– connection to distribution board 314-4
– control of 314-2
– cookers Appx. 5
– disconnecting times, for earthing and automatic disconnection (*See Disconnecting times*)
– earth fault loop impedances (*See Earth fault loop impedances*)
– number of points supplied 314-3
– number required 314-3
– protective conductor impedances Appx. 7
– radial, with BS 1363 socket outlets Appx.5
– ring–
  – continuity test 613-2
  – definition p. 11
  – protective conductor of 543-6, 613-2
  – with BS 1363 socket outlets Appx. 5
– separate control of 314-2
– separation of 314-4
– standard arrangements of 314-3, Appx. 5
– with BS 196 socket outlets Appx. 5
– with BS 4343 socket outlets Appx. 5
Fire alarm systems, applicability of the Regulations to 11-3 (iii)
Fire alarm systems, segregation from LV circuits 525-1 to 525-9
Fire authority, siting of fireman's switch 476-13
Fire barriers 528-1

Fi–Gu

Fire, protection against– Sec. 422
– appliances, fixed 422-2
– distribution boards, backless 422-3
– equipment containing flammable liquid 422-5
– fire hazard to building materials 422-1, 422-2
– fire-resistant shield or enclosure 422-2
– fundamental requirement 13-18
– heat dissipation of equipment 422-1
– heating cables 554-31
– lamps 422-2, 422-4
– luminaires 422-2, 422-4
– surface temperature of equipment 422-2
– ventilation of equipment 422-2
Fireman's switch, provision of 476-12, 476-13
Fireman's switch, selection of 537-17
Fire-resistant–
– shields or enclosures for equipment 422-2
– structural elements, cables passing through 528-1
Fire risks, special, the Regulations not applicable to 11-3 (ii)
Fixed equipment–
– definition p. 9
– disconnecting times for 413-4
– earth fault loop impedances for 413-5
– heat dissipation 422-1
Flammable dielectric liquid, in equipment 422-5
Flexible cables and cords (*See Cables, flexible*)
Flexible conduits, prohibited as protective conductors 543-11
Floor-warming systems–
– cables and conductors for 554-31 to 554-34
– cables with 523-2
– diversity Appx. 4 (Table 4B)
Floors–
– cables in, corrosion Appx. 10
– cables passing through 528-1
– cables under 523-20
– conductive, in earth-free bonded location 413-34
– heated, cables in 523-2
– insulating, resistance of 413-29
Fluctuating loads, assessment for compatibility 331-1
Fluorescent lighting loads, assessment for compatibility of harmonics 331-1
Fluorescent lighting (*See also Discharge lighting*)
Foreign Standards Preface p. vi
Foundation earthing 542-10
Frequency–
– effect on current-carrying capacity of conductors Appx. 9 (Preface)
– suitability of equipment for 512-3
– supply, assessment of 313-1
Functional earthing–
– assessment of need for 331-1
– definition p. 10
– protective earthing in relation to 542-6
Functional extra-low voltage (*See FELV systems*)
Functional switching, devices for 537-18, 537-19
Fundamental requirements for safety Chap. 13
Fuse element, definition p. 10
Fuse elements, for semi-enclosed fuses 533-4
Fuse link, definition p. 10
Fuse links–
– marking of intended type 533-2
– non-interchangeability of 533-2 (ii)
– replaceable by unskilled persons 533-1 Note, 533-2
Fused–
– connection units Appx. 5
– plugs 553-1 (iii), 553-3, 553-5 (i), Appx. 5
– spurs Appx.5
Fuses– (*See also Overcurrent protective devices*)
– accessible to unskilled persons 533-1 Note, 533-2
– cartridge type, preferred 533-4
– in plugs 553-1 (iii), 553-3, 553-5 (i)
– marking of 533-1, 533-2
– nominal current to be indicated 533-1
– non-interchangeability of links 533-2 (ii)
– overload and short circuit protection by 432-2, 533-2
– prohibited in earthed neutral 13-12, 13-13
– removable whilst energised 533-3
– selection of 533-1 to 533-4
– semi-enclosed–
– elements for 533-4
– overload protection by 433-2, Appx. 9 (Preface, 4)
– special ambient temperature correction factors for cables protected by Appx. 9 (Preface, 4)
– shock protection by 413-5, 413-9, 413-12
– short-circuit protection by 432-4
– voltage rating for fault conditions 530-3
Fusing factor 533-2 (ii)

G

Gas meters 547-3
Gas pipes–
– bonding at points of contact with 525-10
– main bonding of 413-2, 547-1 to 547-3
– public, prohibited as earth electrodes 542-14
– segregation from 525-10, 525-11
General characteristics of installation, assessment of 300-1
Glands, cable 527-8
Glossary (*See Definitions*)
Guarding of equipment, against burns 423-1
Guarding of lamps and luminaires 422-4

H

Harmonic currents, assessment for compatibility 331-1
Harmonization Documents Preface p. vi
Health and Safety at Work etc. Act Appx. 2
Heat dissipation of equipment 422-1
Heating appliances–
– bathrooms 471-39
– diversity Appx. 4 (Table 4B)
– standard final circuits and Appx. 5
Heating cables 554-31 to 554-34
Heating equipment, mechanical maintenance switching for 476-7
High-frequency oscillations, assessment for compatibility 331-1
High voltage discharge lighting– 554-3 to 554-19
– ancillary equipment 554-5
– autotransformers 554-6
– bare or lightly insulated conductors 554-9, 554-10, 554-11
– cables 554-8, 554-11, 554-15 to 554-18
– capacitors 554-5
– creepages and clearances 554-12, 554-13
– earth leakage 554-4
– earthed return conductors 554-19
– fireman's switch 476-12, 476-13, 537-17
– inductors 554-5
– isolation 554-6, 554-7
– resistors 554-5
– selection and erection 554-3 to 554-19
– separation from mains supply 554-6
– short-circuit protection 554-4
– supports for cables and conductors 554-11, 554-14
– switching 554-7
– transformers 554-4 to 554-6
– voltage limit 554-3, 554-6
– warning notices 554-5, 554-17
High voltage electrode water heaters and boilers 554-23
Highly Flammable Liquids and Liquified Petroleum Gases Regulations Appx. 2
Hoist shaft, cables in 525-12
Household or similar installations–
– cookers, current demand Appx. 4
– diversity in Appx. 4 (Table 4B)
– final circuits Appx. 5
– socket-outlet circuits in TT systems 471-13
– socket outlets for, selection 553-4
Hydrocarbons, exposure of non-metallic wiring system to 523-17

I

Immersion heaters and standard final circuits Appx. 5
Immersion heaters (*See also Water heaters*)
Impedance–
– circuit, for automatic disconnection 413-3
– earth fault loop (*See Earth fault loop impedance*)
– earthing, in IT systems 413-13 Note
– protective conductor, in socket outlet circuits Appx. 7
– 'protective' (*See Limitation of discharge of energy*)
Indicators–
– circuit breaker overcurrent settings 533-5
– isolating devices, operation of 537-5
– mechanical maintenance switching 537-9
– switchgear, operation of remote 514-1
Indirect contact, definition p. 10
Indirect contact, protection against– 410-1, Secs. 413, 471
– application of 471-11 to 471-46
– automatic disconnection of supply 413-2 to 413-17, 471-11 to 471-15
– Class II equipment or equivalent 413-18 to 413-26, 471-16 to 471-18
– earth-free local equipotential bonding 413-32 to 413-34, 471-20
– earthed equipotential bonding and automatic disconnection 413-2 to 413-17, 471-11 to 471-15
– electrical separation 413-35 to 413-39, 471-21
– exemptions, special 471-26
– non-conducting location 413-27 to 413-31, 471-19
– provision of Sec. 413
– testing 613-2 to 613-4, 613-10, 613-13, 613-15, 613-16
Inductance, mutual, assessment for compatibility 331-1
Inductive equipment, switches for 512-1 Note 1, 512-2 Note
Inductors for discharge lighting 551-4, 554-5
Industrial plugs and socket outlets 553-3
Inlet, caravan 553-13
Inspection–
– certificate 614-1 Notes, 631-1, Appx. 16
– equipment to be accessible for 13-2
– initial, check list Appx. 14
– initial, requirement for Sec. 612
– installation, fundamental requirement 13-20
– periodic–
  – assessment of need for 341-1
  – lengths of periods Appx. 16 (footnotes)
  – notice on 514-5
  – requirement for Chap. 63
– visual, check list Appx. 14
– visual, requirement for Sec. 612

Installation, electrical, definition p. 9
Instantaneous water heaters in bathrooms 471-39
Instantaneous water heaters (*See also Water heaters*)
Instructed person, definition p. 10
Instructed persons–
– areas reserved for, to be indicated by signs 471-25
– emergency switching of safety services reserved for 476-10
– obstacles permissible for protection of 471-9, 471-22
– placing out of reach permissible for protection of 471-10, 471-22
Insulating floor (or wall), definition p. 10
Insulation–
– applied during erection 412-2 Note 2, 613-9
– basic, definition p. 7
– definition p. 10
– double, definition p. 8
– double, of equipment 413-18 to 413-26
– equivalent to Class II (*See Class II equipment*)
– FELV circuits 411-12, 411-13
– joints in conductors 527-3
– live parts 412-1, 412-2, 471-6
– monitoring, in IT systems 413-16
– non-conducting floors and walls 413-29, 613-13
– PEN conductors 546-7
– protection by insulation of live parts 412-1, 471-6
– protective conductors 543-15, 544-4
– reinforced–
– applied during erection 413-18 (iii)
– definition p. 11
– of equipment 413-18 to 413-26
– resistance, testing 613-5 to 613-8
– SELV circuits 411-10
– site-built assemblies 412-2 Note 2, 413-18 (ii), 471-16, 613-9, 613-10
– supplementary, applied during erection 413-18 (ii), 471-16, 613-10
– supplementary, definition p. 12
– thermal, cables in 522-6
Interlocking of enclosure opening 412-6
Interlocking of isolating devices 461-3
International Electrotechnical Commission Preface p. vi, Chap. 32, Appx. 6
Intruder alarm systems, applicability of the Regulations to 11-3 (iii)
Isolation– Secs. 460 & 461, 476-2 to 476-6
– at origin of installation 13-14, 476-2
– definition p. 10
– devices for, contact distances 537-3
– devices for, selection and erection Sec. 537
– discharge lighting 476-6, 554-6
– fundamental requirements 13-14
– HV discharge lighting circuits 554-6

Isolation – *continued*
– indication of 537-5
– main, at origin of installation 476-2
– motor circuits 476-5
– motors, fundamental requirement 13-15
– multipole devices preferred 537-7
– off-load, securing against inadvertent operation 537-6
– PEN conductor not to be isolated 460-2, 546-6
– prevention of inadvertent reclosure 537-6
– remote devices for 476-4
– semiconductors unsuitable for 537-4
– single-pole devices for 537-7
Isolator–
– contact distances 537-3
– for maintenance of main switchgear 476-3
– off-load, securing against inadvertent operation 537-6
– semiconductors unsuitable as 537-4
IT system–
– autotransformers, step-up, prohibited 551-2
– connection of installation to Earth 542-4
– definition p. 12
– distribution of neutral in 473-13, 512-1
– earthing of live parts in 413-13
– earthing resistance 542-7
– explanatory notes on Appx. 3
– exposed conductive parts in, connection of 413-10, 413-11
– first fault in 413-16
– insulation monitoring devices for 413-16
– neutral distribution in 473-13, 512-1
– overload protection 473-4
– overvoltage in 413-13 Note
– prohibited for public supplies in UK Appx. 3
– shock protective devices in 413-15, 471-15
– variation of earthing resistance 531-5
– voltage oscillations in 413-13 Note

J

Joints–
– cable, accessibility of 526-1
– cable and conductor Sec. 527
– conduit system, accessibility of 526-2
– earthing arrangements 542-20
– fundamental requirements 13-6
– protective conductor 543-16, 543-17
Joists, cables passing 523-20

K

Keys–
– adjustment of over-current settings by 533-5
– areas accessible only to skilled persons by 471-24
– opening enclosures by 412-6, 413-24
Kitchens and ring final circuits Appx. 5

L

Labels– (*See also Marking, Warning notices*)
– earth electrode connections 542-18
– HV discharge lighting cables 554-17
– switchgear and controlgear 514-1
Lacquers, as insulation 412-2 Note 2, 413-21 Note
Lampholders–
– as example of permissible opening in enclosure 471-8
– bathrooms 471-38
– batten, defined as luminaire p. 10
– centre contact 553-18
– current demand of Appx. 4
– Edison type screw 553-14, 553-18
– filament lamp, voltage limit 553-15
– overcurrent protection of 553-14
– pendant, defined as luminaire p. 10
– selection and erection of 553-14 to 553-18
– temperature rating 553-16
Lamps, fire hazard from 422-2, 422-4
Leakage currents and residual current devices 531-5
Leakage currents (*See also Earth leakage*)
Licensing, premises subject to 12-7, Appx. 2
Lids in 'Class II' enclosures 413-24
Life of installation, assessment of 341-1
Lift shafts, cables in 525-12
Lifting magnets, omission of overload protection 473-3
Lighting– (*See also Luminaires*)
– diversity Appx. 4 (Table 4B)
– outlet, current demand of Appx. 4
Lightning protection, the Regulations not applicable to 11-3 (xi)
Lightning protection, earthing for 541-2 Note
Limitation of discharge of energy, protection by 411-16, 471-5
Live conductors, assessment of types 312-2
Live conductors, earthing of, prohibited in IT systems 413-13
Live part, definition p. 10
Live parts–
– bare, in SELV circuits 411-10
– bare, placing out of reach 412-10
– of SELV circuits, separation of 411-4, 411-6
Livestock–
– protective measures in situations for 471-40, 471-41
– wiring systems in locations for 523-34
Local supplementary bonding 413-7, 471-35
Low voltage–
– definition p. 12
– range 11-2
– reduced 471-27 to 471-33
Luminaire, definition p. 10
Luminaire track system (*See definition of Socket outlet, p. 11*)
Luminaires–
– bathrooms 471-38
– caravans 554-2
– fire hazard from 422-2, 422-4
– flexible cords for 521-5, 521-6
– pendant–
 – in caravans 554-2
 – ceiling rose for 553-20
 – mass suspended 523-32, 554-1
– portable, connection of 553-9
– supported by flexible cords 523-32
– suspended from non-metallic boxes 523-5
– switches mounted on 476-8, 476-18
– switching for 476-17 to 476-20

M

Machines– (*See also Motors*)
– emergency switching for 476-11
– rotating, exciter circuits of 473-3
– rotating, selection and erection of Sec. 552
Magnetic–
– circuit of residual current device 531-4
– equipment, mechanical maintenance switching for 476-7
– fields, effect on residual current devices 531-7
Magnets, lifting, omission of overload protection 473-5
Main earthing terminal–
– connection to Earth 542-1 to 542-9
– definition p. 10
– selection and erection of 542-19, 542-20
Main equipotential bonding–
– conductors, selection and erection 547-2, 547-3
– provision of 413-2
– zone, equipment used outside 471-12
Main switch for installation 476-15
Maintainability, assessment of 341-1

Ma–Ne

Maintenance of equipment–
– assessment of frequency of 341-1
– fundamental requirement 13-2
Markets, closed, discharge lighting in 476-12
Marking– (*See also Labels, Warning notices*)
– buried cables 523-23
– caravan inlets 553-13
– emergency switching devices 463-4
– fuses and circuit breakers 533-1, 533-2
– isolating devices 461-5
– mechanical maintenance switching devices 462-2
Mass supported by flexible cords 523-32
Mass suspended from non-metallic boxes 523-5
Materials–
– established, use of 12-5
– new, use of 12-6
– proper, use of 13-1
Maximum demand, assessment of 311-1, Appx. 4
Maximum demand, suitability of supply for 313-1
Mechanical maintenance, definition p. 10
Mechanical maintenance switching, requirements Sec. 462, 476-7, 476-8
Mechanical maintenance switching devices, selection of 537-8 to 537-11
Mechanical stresses–
– cables and conductors 434-1, 523-19 to 523-33
– earthing arrangements 542-7, 542-16
– earthing conductors 542-16
– heating cables and conductors 554-31 to 554-33
– plugs and socket outlets 553-8
– protective conductors 543-1, 543-15, 547-4 to 547-6
– short circuit 434-1
Metalliferous mine installations Appx. 2
Metalwork– (*See also Exposed conductive parts, Extraneous conductive parts*)
– bonding of 413-2, 413-7
– cables passing through 523-21
– exposed, of safety separated circuits 413-38, 413-39
– of other public services, prohibited as earth electrode 542-14
– wiring system, as protective conductor 543-9
– wiring system, protection against corrosion 523-17, Appx. 10
Mines and quarries, installations at–
– statutory regulations for Appx. 2 (1)
– the Regulations not applicable to 11-3 (ix)
Miniature circuit breakers (*See Circuit breakers*)
Mobile equipment in non-conducting location 413-30
Mobile safety sources–
– electrically separated circuits supplied by 413-36
– SELV circuits supplied by 411-9
Monitoring systems, earthing 543-18
Monitoring systems, insulation, IT systems 413-16
Motor vehicles, applicability of the Regulations to 11-3 (v)
Motors–
– automatic restarting 552-4
– control of 552-3, 552-4
– disconnection, fundamental requirement 13-15
– diversity Appx. 4 (Table 4B)
– frequent start 552-1
– isolation of 476-5
– mechanical maintenance switching 476-7
– stalled, capability of emergency switching for 537-12
– starters, autotransformer 551-1 Note
– starters, coordination of overload and short-circuit protection 435-1 Note
– starting currents 552-1, 552-2
– starting, voltage drop 522-8
Mutual detrimental influence–
– different types of equipment 515-2
– electrical and other services 515-1, 525-10 to 525-12
– LV circuits and other circuits 525-1 to 525-9
– protective measures 470-2
Mutual inductance, assessment for compatibility 331-1

N

Neutral–
– conductor–
  – combined with protective conductor (*See PEN conductors*)
  – definition p. 10
  – earthing of 13-13 Note, 471-30, 546-1, 554-24, Appx. 3
  – fuse prohibited in 13-12, 553-1 (iii)
  – identification Sec. 524
  – isolation of 460-2
  – minimum cross-sectional area 522-9
  – overcurrent detection in 473-10 to 473-12
  – reduced, prohibited in discharge lighting circuits 522-9
  – single-pole devices prohibited in 13-13, 530-2
  – switching of 530-1, 530-2
– distribution of, in IT systems 473-13, 473-14, 512-1
– earthing of 13-13 Note, 471-30, 546-1, 554-24, 554-25, Appx. 3
– links 476-15, 537-7
– points, artificial, in IT systems 413-13 Note
– reduced, prohibited in discharge lighting circuits 522-9

New materials and methods, use of 12-5
Nominal current of fuses and circuit breakers 533-1
Nominal voltage–
– assessment 313-1
– definition p. 10
– reduced low voltage circuits 471-28
– SELV circuits 411-2, 411-10, 471-2, 471-3, 471-41
Non-conducting location, protection by– 413-27 to 413-31, 471-19
– application of 471-19
– basic requirements 413-27
– limited to special situations 471-19
– mobile and portable equipment in 413-30
– not recognised for general use 471-19
– permanency of arrangements 413-30
– precautions against propagation of potential outside location 413-31
– protective conductors prohibited 413-28
– resistance of floors and walls 413-29, 613-13
– socket outlets in 413-28
Notes to the Regulations, status of 12-8
Notices– Sec. 514
– fireman's switch 476-13, 537-17
– periodic inspection and testing 514-5
– socket outlets for equipment used outdoors 514-8
– warning–
– caravans 514-6
– earthing and bonding conductors 514-7, 542-18
– emergency switching 537-16
– isolation of equipment 461-3
– HV discharge lighting 554-5, 554-17
– voltages exceeding 250V 514-4
Numbering system of Regulations p. vii

O

Object and effects of the Regulations Chap. 12
Obstacle, definition p. 10
Obstacles, protection by– 412-7, 412-8, 471-9
– application of 471-9
– basic requirements 412-7, 412-8
– limited to areas for skilled or instructed persons only 471-9, 471-22
Offshore installations, Regulations not applicable to 11-3 (vii)
Oil–
– exposure of non-metallic wiring systems to 523-17
– filled equipment (*See Flammable dielectric*)
– services, bonding at points of contact 525-10
– services, segregation from 525-10, 525-11
Operational conditions, cables and conductors Sec. 522
Operational conditions, equipment Sec. 512
Origin of an installation–
– definition p. 10
– isolation at 13-14, 476-2
– overcurrent protection at 13-7 Note
Oscillations, HF, assessment for compatibility 331-1
Oscillations, voltage, in IT systems 413-13 Note
Other services, bonding of 13-11, 413-2
Overcurrent, definition p. 10
Overcurrent detection–
– definition p. 10
– neutral conductor, IT systems 473-13, 473-14
– neutral conductor, TN or TT systems 473-10 to 473-12
– phase conductors 473-9
Overcurrent, limitation by supply characteristics 436-1
Overcurrent, protection against– Chap. 43, Sec. 473
– at origin of installation 13-7 Note
– basic requirement 13-7, 431-1
– discrimination in 533-6
– electrode water heaters and boilers 554-21, 554-26
– fundamental requirement 13-7
– lampholders 553-14
– limitation by supply characteristics 436-1
– motors 552-3
– neutral conductor, IT systems 473-13 to 473-14
– neutral conductor, TN or TT systems 473-10 to 473-12
– paralleled conductors 433-3, 434-7
– phase conductors 473-9
– settings of adjustable circuit breakers 533-5
Overcurrent protective devices–
– at origin, assessment of 313-1
– at origin, use of supply undertaking device 13-7
– coordination of characteristics, for overload and short-circuit protection 435-1
– discrimination 533-6
– fundamental requirements 13-7
– overload and short-circuit protection 432-2
– overload protection only 432-3
– selection and erection of Sec. 533
– shock protection, as 413-5, 413-9, 413-10, 413-12
– short-circuit protection only 432-4
– supply undertaking's, use of 13-7
Overcurrent settings of circuit breakers 533-5
Overhead Line Regulations 412-9, Appx. 11 (16)
Overhead lines–
– insulator wall brackets, exemption from protection against indirect contact 471-26 (i)
– placing out of reach 412-9
– supports Appx. 11 (16)
– types of conductor 521-4

Ov–Po

Overhead wiring supports Appx. 11 (13, 14, 15, 17)
Overhead wiring systems, selection and erection 523-26
Overheating of switchgear 421-1
Overload–
– alarm 473-3 Note
– current, definition p. 10
– protection against– Sec. 433, 473-1 to 473-4
  – application of 473-1 to 473-4
  – coordination with short-circuit protection 435-1
  – omission of 473-3
  – paralleled conductors 433-3
  – position of devices for 473-1, 473-2
– protective devices 432-1, 432-3, 433-1, 433-2, 473-4
Overvoltages, in IT systems 413-13 Note
Overvoltages, transient, assessment for compatibility 331-1

P

Paints, as insulation 412-2 Note 2, 413-21 Note
Paralleled cables, current-carrying capacity 522-3
Paralleled conductors, overload protection 433-3
Paralleled conductors, short-circuit protection 434-7
Partitions in trunking 525-5 to 525-7
Passageways for open type switchboards 471-23
PEN conductor, definition p. 10
PEN conductors–
– cable enclosure prohibited as 543-13
– isolation or switching of, prohibited 460-2
– residual current device with, prohibited 413-9
– selection and erection of Sec. 546
Pendant lampholder, defined as luminaire p. 10
Pendant luminaire–
– caravans 544-2
– ceiling rose for 553-20
– mass suspended 554-1
Periodic inspection–
– assessment of need for 341-1
– length of periods Appx. 16 (footnotes)
– notice on 514-5
Persons–
– instructed, definition p. 10
– skilled, definition p. 11
– skilled or instructed, obstacles permissible for protection of 471-9, 471-22
– skilled or instructed, placing out of reach permissible for protection of 471-10, 471-22
– untrained, the Regulations not intended for 12-2
Petroleum (Consolidation) Act Appx. 2
Phase conductor–
– definition p. 11
– identification of Sec. 524
– overcurrent detection in 473-9
Phase, loss of 473-9
Pipes–
– colour identification of 524-2
– gas (*See Gas pipes*)
– earth electrodes formed by 542-10, 542-14
– underground, for cables 523-24
– water (*See Water pipes*)
Placing out of reach, protection by– 412-9 to 412-13, 471-10
– application of 471-10
– bare live parts 412-10 to 412-12
– limited to locations for skilled or instructed persons only 471-10, 471-22
– overhead lines 412-9
– with obstacles 412-12
– zone of accessibility 412-10, 412-12, 412-13
Plan of the 15th Edition, Notes on p. vii
Plasters, wiring systems in contact with Appx. 10
Platforms, working, for open type switchboards 471-23
Plug, definition p. 11
Plugs–
– caravans 553-6
– clock 553-5
– construction sites 553-7
– emergency switching by, prohibited 537-13
– FELV circuits 411-15
– functional switching by 537-18
– fused, selection of 553-1 (iii), 553-3, 553-5 (i), Appx. 5
– 'instantaneous' water heaters not to be supplied by 554-30
– reduced low voltage circuits 471-33
– selection and erection of 553-1 to 553-9
– SELV circuits 411-8
– shavers 553-5 (ii)
– special circuits 553-5 (iii)
PME–
– Approval, 1974 413-2, Appx. 2 (4), Appx. 3
– on caravan sites 471-43
– main bonding connections for 547-2
Point (in wiring), definition p. 11
Points, current demand of Appx. 4
Polarity, test of 613-14
Poles, concrete, exemption from protection against indirect contact 471-26 (ii)
Polluting substances, wiring systems exposed to 523-17
Portable equipment–
– connection of 553-9
– definition p. 11
– in non-conducting location 413-30
– prohibited in bathrooms 471-34
Post Office earth wires, not to be bonded 413-2 Note 3, 525-10 Note 2
Potentially explosive atmospheres, installations in, extent covered by Regulations 11-3 (ii)

Po–Re

Power demand, suitability of equipment for 13-3, 512-4
Power factor, discharge lighting circuits Appx. 4 (Table 4A Note)
Power factor, effect upon voltage drop Appx. 9 (Preface, 1)
Prefabricated equipment, applicability of the Regulations to 11-4
Premises subject to licensing 12-7
Proper materials, use of 13-1
Prospective short-circuit current–
– at origin, assessment of 313-1
– at points of protective devices Sec. 432
– determination of 434-2
Protection against electric shock–
– application of measures Sec. 471
– protective measures Chap. 41
Protection against overcurrent–
– application of measures Sec. 473
– protective measures Chap. 43
Protection against thermal effects Chap. 42
Protection for safety Part 4
Protective conductor, definition p. 11
Protective conductors–
– accessibility of connections 543-16
– between separate installations 542-9
– bonding, selection and erection of Sec. 547
– caravan sites 471-43
– 'Class II' equipment in relation to 413-25, 471-17
– colour identification of 524-1, 524-3, 524-5
– combined with neutral conductors (*See PEN*)
– continuity of 543-15 to 543-19
– cross-sectional areas 543-1 to 543-3, 546-2, Sec. 547
– fault-voltage operated devices 544-4, 544-5
– identification of 524-1, 524-3 to 524-5
– impedances of, for socket outlet circuits Appx. 7
– insulation of 543-15, 544-4
– non-conducting location, prohibited in 413-28
– preservation of continuity 543-15 to 543-19
– residual current device, to be outside magnetic circuit of 531-4
– ring final circuit 543-6
– safety separated circuits 413-39
– selection and erection of Secs. 543 to 547
– switching prohibited in 543-17
– testing of 613-2, 613-3, Appx. 15 (3)
– types of, description 543-4 to 543-14, 547-7
– types of, example Appx. 13
Protective devices and switches, position of 13-12, 13-13
Protective devices and switches, identification of 514-2
Protective devices (*See also Overcurrent protective devices, Residual current devices, Fault-voltage operated devices, Monitoring systems*)
'Protective impedance' (*See Limitation of discharge of energy*)
Protective Multiple Earthing (*See PME*)
Public supply systems–
– effect of starting currents on 331-1 Note
– IT system prohibited for (in UK) Appx. 3
– switching arrangements for safety and standby supplies 313-2
– the Regulations not applicable to 11-3 (i)

Q

Quarry installations Appx. 2 (1)

R

Radiators, bonding of 413-7 Note
Radial final circuits, with BS 1363 socket outlets Appx. 5
Radio equipment and circuits, applicability of the Regulations to 11-3 (iii)
Radio interference suppression equipment, applicability of the Regulations to 11-3 (x)
Rapidly fluctuating loads, assessment for compatibility 331-1
Reduced low voltage systems 471-27 to 471-33
Regulations, Statutory–
– list of Appx. 2
– relationship with the Regulations 12-3, Appx. 2
Regulations, The–
– amendments to p. vi
– citation in contracts 12-2
– date of validity 12-9
– departures from 12-3, 12-5, Appx. 2
– editions, list of p.iv
– effect of 12-3
– exclusions from scope of 11-3
– not intended to instruct untrained persons 12-2
– not to be regarded as specification 12-2
– notes to, status of 12-8
– objects of Chap. 12
– Parts 3 to 6, relationship to Chapter 13 12-3
– relationship to statutory regulations 12-3, Appx. 2
– scope of 11-1
– voltage ranges dealt with 11-2
Reinforced concrete (*See Steel*)
Reinforced insulation, definition p. 11

Re–Se

Reinforced insulation, of equipment 413-18 to 413-26
Reliability of equipment for intended life 341-1
Remote devices for isolation 476-4
Remote switching for mechanical maintenance 537-15, 537-16
Repairs, notification of need for 622-1
Residual current device, definition p. 11
Residual current devices–
– application of, as shock protection 471-14
– auxiliary supply for 531-6
– caravan installations 471-43, 471-44
– conductors to be controlled by 531-1
– direct contact protection by, not recognised as sole method 471-14 Note
– earth fault loop impedance for 413-6
– fundamental requirement for 13-10
– HV electrode water heaters and boilers 554-23
– magnetic circuit of 531-4
– magnetic fields of other equipment 531-7
– operating current (*See Residual operating current*)
– PEN conductor circuits, prohibited in 413-9, 471-14, 546-2
– phase conductors to be controlled by 531-1
– preferred, for shock protection in TT systems 413-12 Note
– preferred, where shock protection by overcurrent devices impracticable 471-14
– prohibited in circuits incorporating PEN conductor 413-9, 471-14, 546-2
– protection against direct contact by, not recognised as sole method 471-14 Note
– reduced low voltage circuits, with 471-32
– required, for caravan sites 471-44
– required, for socket outlet circuits in household TT systems 471-13
– required, for socket outlet circuits supplying equipment outside main equipotential zone 471-12
– selection and erection of 531-3 to 531-8
– shock protection by–
– earth fault loop impedance 413-6
– in IT systems 413-15
– in TN systems 413-9
– in TT systems 413-10, 413-12
– in TT systems, preferred 413-12 Note
– short-circuit capacity 531-8
– supplementary protection against direct contact by 471-14 Note
– testing 613-16, Appx. 15(6)
Residual operating current–
– caravan installations 471-44
– definition p. 11
– earth fault loop impedance in relation to 413-6
Residual operating current – *continued*
– equipment outside main equipotential zone 471-12
– IT system overcurrent protection, for 473-14
– reduced LV circuit device, for 471-32
– selection of 531-5
– socket outlet circuits in household TT systems 471-13
Resistance area (of an earth electrode), definition p. 11
Resistance of insulating floors and walls 413-29
Resistance of human body (*See Body resistance*)
Resistors for HV discharge lighting 554-5
Ring final circuit–
– circuit protective conductor of 543-6
– definition p. 11
– standard, with BS 1363 socket outlets Appx. 5
– test of continuity 613-2, Appx. 15(2)
Road-warming cables and conductors 554-31 to 554-34
Roads, overhead lines crossing Appx. 11
Rotating machines– (*See also Motors*)
– emergency switching for 476-11
– exciter circuits of 473-3
– selection and erection of Sec. 552

S

Safety extra-low voltage, protection by (*See SELV circuits*)
Safety, fundamental requirements for Chap. 13
Safety, object of the Regulations Chap. 12
Safety services, emergency switching of 476-10
Safety services, supplies for 313-2
Safety sources for SELV circuits 411-3
Schedule (*See Diagrams*)
Scope of the Regulations Chap. 11
Scotland, application of the Regulations in Appx. 2
Screws, fixing, for non-metallic accessories 471-26 (iv)
Screws, insulating, not to be relied upon as Class II 413-23
Segregation–
– assemblies of equipment 515-2
– circuits 525-1 to 525-9
– non-electrical services 525-10 to 525-12
Selection and erection of equipment Part 5
SELV circuits–
– agricultural installations 471-41
– application as protective measure 471-2, 471-3
– arrangement of 411-4
– cable couplers for 553-10
– exposed conductive parts of 411-5

Se–So

SELV circuits – *continued*

– live parts of 411-4, 411-6, 411-10
– mobile safety sources for 411-9
– nominal voltage 411-2, 411-10, 471-2, 471-3
– plugs and socket outlets for 411-8, 553-1 (i), 553-2
– safety sources for 411-3, 411-9
– separation from other circuits 411-7

Semiconductors– (*See also Electronic devices*)
– unsuitable as isolators 537-4

Semi-enclosed fuses (*See Fuses*)

Separation, electrical, as protection against shock (*See Electrical separation*)

Shaver supply units, current demand of Appx. 4

Shaver supply units in bathrooms 471-34, 471-37, 471-39

Shavers, plugs and socket outlets for 553-5 (ii)

Shields, fire-resistant, for fixed equipment 422-2

Shields, lampholder, in bathrooms 471-38

Ships, electrical equipment on board, the Regulations not applicable to 11-3 (vi)

Shock current, definition p. 11

Shock, electric, definition p. 9

Shock, electric, protection against (*See Direct contact, Indirect contact*)

Short-circuit current, definition p. 11

Short-circuit current, prospective–
– assessment at origin of installation 313-3
– at points of protective devices Sec. 432
– determination of 434-2
– for residual current devices 531-8

Short circuit, protection against– Sec. 434, 473-5 to 473-8
– devices for 432-1, 432-4, 434-4 to 434-6
– HV discharge lighting circuits 554-4
– omission of devices for 473-8
– position of devices for 473-5 to 473-7

Showers, rooms containing, protective measures 471-34 to 471-39

Signal of first fault in IT system 413-16

Signs, electric (*See Discharge lighting*)

Signs, warning, for areas accessible only to skilled or instructed persons 471-25

Signs, warning (*See also Warning notices*)

Simultaneously accessible parts–
– bathrooms 471-35, 471-36
– definition p. 11

Simultaneously accessible parts – *continued*

– earth-free local bonding locations 413-32
– SELV circuits 471-2
– with automatic disconnection 413-3

Sinks, bonding of 413-7 Note

Site-built assemblies, insulation of–
– as protection against direct contact 412-2 Notes, 613-9
– as protection against indirect contact 413-18 (ii), 471-16, 613-10
– testing 613-9, 613-10

Skilled person, definition p. 11

Skilled persons–
– areas reserved for, to be indicated by signs 471-25
– emergency switching of safety services by 476-10
– exemption from requirements for shock protection, in locked areas for 471-24
– obstacles permissible for protection of 471-9, 471-22
– placing out of reach permissible for protection of 471-10, 471-22

Socket outlet, definition p. 11

Socket outlets–
– bathrooms, prohibited in 471-34
– bonding to boxes etc. 543-10
– caravan site installations 471-43, 471-44, 553-6
– caravans, in 471-45
– circuits for–
  – disconnecting times 413-4
  – earth fault loop impedances for 413-5
  – household TT systems 471-13
  – protective conductor impedances for Appx. 7
  – standard arrangements of Appx. 5
  – supplying outdoor equipment 471-12
– clocks, for 553-5
– construction sites 553-7
– current demand of Appx. 4
– diversity Appx. 4 (Table 4B)
– emergency switching by, prohibited 537-13
– FELV circuits 411-15
– final circuits for Appx. 5
– functional switching by 537-18
– height of 553-8
– household installations, selection 553-4
– non-conducting location, in 413-28
– outdoor equipment, for 471-12
– prohibited–
  – bathrooms 471-34
  – 'Class II' circuits or installations 471-18
  – emergency switching 537-13
  – 'instantaneous' water heaters 554-30
– provision of 553-9
– reduced low voltage circuits 471-33
– safety separated circuits 413-39
– selection and erection of 553-1 to 553-9, Appx. 5
– SELV circuits 411-8, 553-1 (i)

So–Sw

Socket outlets – *continued*
– shavers, for 553-5 (ii)
– supplying equipment outside main equipotential zone 471-12
Soil warming, cables and conductors for 554-31 to 554-34
Soldered joints, corrosive fluxes 523-18
Soldered joints, short-circuit withstand 434-6
Sound distribution circuits, applicability of the Regulations to 11-3 (iii)
Space factor, definition p. 11
Space factor, in conduit and trunking 529-7, Appx.12
Space heating, diversity Appx. 4 (Table 4B)
Special installations needing qualified advice 12-2
Special situations–
– protection by non-conducting location 471-19
– protection by earth-free local equipotential bonding 471-20
– protection by electrical separation 471-21
Specification, the Regulations not intended as 12-2
Specifications (*See Standards*)
Spur, definition p. 12
Spurs in standard final circuits Appx. 5
Standard circuit arrangements–
– details of Appx. 5
– diversity not applicable to Appx. 4
Standard methods of testing Appx. 15
Standards–
– British Preface p. vi, 511-1, 612-1
– British, mentioned in the Regulations, list Appx. 1
– foreign Preface p. vi
Standby supplies 313-2
Starters, discharge lighting (*See Discharge lighting*)
Starters, motor (*See Motors*)
Starting currents, assessment for compatibility 331-1
Stationary equipment, current demand of Appx. 4
Stationary equipment, definition p. 12
Statutory regulations (*See Regulations, statutory*)
Steel–
– reinforced concrete poles, exemption from protection against indirect contact 471-26 (ii)
– reinforcement of concrete, use as earth electrodes 542-10
– structural metalwork, bonding of 413-2, 413-7
Storage heaters, diversity Appx. 4 (Table 4B)
Structural metalwork, bonding of 413-2, 413-7
Sunlight, cables exposed to 523-35
Supplementary bonding conductors, selection and erection of 547-4 to 547-7
Supplementary equipotential bonding (*See Equipotential bonding*)
Supplementary insulation, applied during erection 413-18 (ii), 471-16, 613-10
Supplementary insulation, definition p. 12
Supplies–
– electrically separated circuits 413-36
– reduced low voltage circuits 471-29
– safety services, assessment 313-2
– SELV circuits 411-3
– standby, assessment 313-2
Supply, nature of, assessment 313-1
Supply systems, public–
– effect of starting currents on 331-1 Note
– IT system prohibited for (in UK) Appx. 3
– switching arrangements for safety and standby supplies 313-2
– the Regulations not applicable to 11-3 (i)
Supply undertakings–
– consultation with, on starting currents 331-1 Note
– disputes with consumers, procedure Appx. 2 (2)
– equipment, suitability for alterations to installation 13-19
– not compelled to give supply in certain circumstances Appx. 2 (2)
– providing part of earth fault current path Appx. 3
– switchgear, use for isolation and protection of installation 13-7 Note, 13-14 Note
Supports for cables for fixed wiring 529-1, Appx. 11
Supports for wiring systems 529-1, 529-2, Appx. 11
Switch, definition p. 12
Switch, linked–
– definition p. 12
– 'instantaneous' water heaters 554-30
– main 476-15
– selection of type 530-1
– step-up transformers 551-3
Switch, main, for installation 476-15
Switchboard, definition p. 12
Switchboards, conductors on 522-1, 522-2, 524-3
Switchboards, open type, working platforms etc. 471-23
Switches–
– appliances, mounted on 476-8
– bathrooms 471-39
– capacitive or inductive equipment, for 512-1 Note 1, 512-2 Note
– discharge lighting 537-19
– fireman's 476-12, 476-13, 537-17
– linked (*See Switch, linked*)
– luminaires, mounted on 476-8

Sw–Te

Switches *continued*
- main switch for installation 476-15
- position of, fundamental requirement 13-13
- prohibited in PEN conductors 460-2, 546-6
- prohibited in protective conductors 543-17
- single-pole, prohibited in neutral 13-13, 530-2
- step-up transformers, for 551-3

Switchgear–
- definition p. 12
- for isolation and switching Sec. 537
- for overcurrent protection Sec. 533
- for shock protection Sec. 531
- indicators for 514-1
- labels for 514-1
- linked, selection of type 530-2
- main, isolator for maintenance of 476-3
- marking of 514-4
- overheating and arcing 421-1
- selection and erection of Chap. 53
- single-pole, prohibited in neutral 13-12, 13-13, 530-2
- use of supply undertaking's 13-14 Note, 13-7 Note

Switching– Chap. 46, Sec. 476
- appliances and luminaires 476-17 to 476-20
- cooking appliances 476-20
- devices, selection and erection of Sec. 537
- electrode water heaters and boilers 554-21
- emergency Sec. 463, 476-9 to 476-13
- fundamental requirements for 13-14, 13-15
- 'instantaneous' water heaters 554-30
- main switch for installation 476-15
- mechanical maintenance Sec. 462, 476-7, 476-8
- of circuits 476-16
- prohibited, in PEN conductor 460-2, 546-6
- prohibited, in protective conductors 543-17
- remote, for mechanical maintenance 537-15, 537-16
- single-pole, prohibited in neutral 13-13, 530-2
- step-up transformers 551-3

Symbols 514-3

System, definition p. 12

T

Tanks and taps, bonding of 413-7 Note

Telecommunications circuits–
- applicability of the Regulations to 11-3 (iii)
- segregation from LV circuits 525-1 to 525-9

Temperature–
- ambient, cables and conductors 523-1 to 523-6, Appx. 9, Appx. 10 (Note 2)
- ambient, definition p. 7
- limiting device for water heaters 554-27
- limits –
  - current-carrying capacity 522-1 to 522-5, Appx. 9
  - heating cables 554-34
  - non-metallic boxes suspending luminaires 523-5
  - protective conductors 543-2
  - short-circuit protection 434-6
  - wiring system enclosures 523-5, 523-6
- low, effect on cables 523-4, Appx. 10 (Note 2)
- rating of lampholders 553-16
- surface, of equipment 422-2, 423-1

Temporary installations–
- cable supports Appx. 11 (12)
- construction sites Appx. 16 (footnotes)
- local authority requirements etc. Appx. 2 (7)
- prefabricated, applicability of the Regulations to 11-5
- the Regulations generally applicable to 11-1

Terminations of cables, conductors, etc. Sec. 527

Testing– Part 6
- alterations to installations Chap. 62
- assessment of need for periodic 341-1
- barriers provided during erection 613-12
- 'Class II' enclosures 413-22
- continuity of protective conductors 613-2, 613-3, Appx. 15 (3)
- continuity of ring circuit conductors 613-2, Appx. 15 (2)
- earth electrode resistance 613-4, Appx. 15 (4)
- earth fault loop impedance 613-15, Appx. 15 (5)
- electrical separation of circuits 613-11
- electronic devices 613-7
- enclosures provided during erection 613-10, 613-12
- equipment, fundamental requirement 13-2
- fault-voltage operated devices 613-16, Appx. 15 (6)
- initial Secs. 611, 613
- installation, fundamental requirement 13-20
- insulation–
  - applied during erection 412-2 Note 2
  - non-conducting floors and walls 613-13
  - resistance 613-5 to 613-8
  - site-built assemblies 613-9, 613-10
- periodic Chap. 63
  - assessment of need for 341-1
  - length of periods Appx. 16 (footnotes)
  - notice on 514-5
- polarity 613-14
- protection against direct contact 613-12
- protective conductors, continuity 613-3, Appx. 15(3)
- repeat 613-1
- residual current devices 613-16, Appx. 15 (6)

Te–Ve

Testing – *continued*

– ring circuit continuity 613-2, Appx. 15 (2)
– test certificates 614-1, Appx. 16
– test voltages 613-5, 613-7

Theatres, plugs and socket outlets for 553-3
Theatres, installations in Appx. 2 (8)
Thermal effects, protection against Chap. 42
Thermal insulation, cables in 522-6
Thermal storage heaters, diversity Appx. 4 (Table 4B)
Thermal stresses of short circuit 434-1
Thermal stresses on earthing arrangements 542-7
Thyristor drives, assessment for compatibility 331-1
Times, disconnecting–
– equipment in bathrooms 471-36
– for compliance with temperature limits for protective conductors Appx. 8
– reduced low voltage circuits 471-32
– shock protection, for fixed equipment outside main equipotential zone 471-12
– shock protection, general 413-4
– short-circuit protection 434-6

TN system–
– connection of exposed conductive parts in 413-8
– definition p. 12
– devices for shock protection 413-9
– earthing resistance 542-7
– explanatory notes on Appx. 3
– switching of neutral 530-1, 530-2

TN-C system–
– connection of installation to Earth 542-5
– definition p. 12
– explanatory notes on Appx. 3
– isolation and switching in 460-2

TN-C-S system–
– connection of installation to Earth 542-3
– definition p. 12
– explanatory notes on Appx. 3
– on caravan site, earthing of 471-43

TN-S system–
– connection of installation to Earth 542-2
– definition p. 12
– earthing on caravan sites 471-43
– explanatory notes on Appx. 3
– isolation of neutral 460-2

Tools, for–
– adjustment of overcurrent settings 533-5
– compression joints 527-6 Note
– disconnection of main earthing terminal 542-20
– disconnection of protective conductors 543-17
– opening areas accessible only to skilled persons 471-24
– opening enclosures 412-6, 413-24

Total insulation of equipment 413-18, 471-16

Track system, luminaire (*See definition of Socket outlet p. 11*)
Traction equipment, the Regulations not applicable to 11-3 (iv)
Transformers–
– bell, current demand of Appx. 4
– current, omission of overload protection 473-3
– HV discharge lighting 554-4 to 554-6
– of residual current devices 531-4
– safety isolating, for SELV circuits 411-3
– selection and erection of Sec. 551

Transient overvoltages, assessment for compatibility 331-1
Trunking (for cables), definition p. 12
Trunking systems–
– cable capacities of 529-7, Appx. 12
– exposed to water or moisture 523-8, 523-15
– fire barriers in 528-1
– joints in 543-16
– partitions in 525-5 to 525-7
– protective conductors formed by 543-7, 543-9, 543-10, 543-13
– selection of 521-12
– socket outlets in 543-10
– space factors in 529-7, Appx. 12
– supports for 529-2, Appx. 11
– terminations 527-5
– vertical runs, temperatures in 523-6

TT system–
– connection of exposed conductive parts in 413-10, 413-11
– connection of installation to Earth 542-4
– definition p. 12
– devices for shock protection 413-12, 471-15
– earthing resistance requirements 542-7
– earthing resistance variations 531-5, 531-9
– explanatory notes on Appx. 3
– socket outlet circuits in household or similar installations 471-13
– switching of neutral 530-1, 530-2

Type-tested equipment, Class II 413-18 (i)

U

Underground cables 523-23, 523-24, Appx. 10

Untrained persons, the Regulations not intended for 12-2

V

Validity of the Regulations, date of 12-9
Varnishes, as insulation 412-2 Note 2, 413-21 Note

Vehicles, cables to be clear of 523-26, Appx. 11 (Table 11B)

Ve–Wo

Vehicles, electrical equipment of, the Regulations not applicable to 11-3 (v)
Ventiliation, fixed equipment 422-2
Ventilation of luminaires in caravans 554-2
Vermin, cables exposed to 523-34
Voltage–
– drop– 522-8, Appx. 9
  – basis of tabulated values Appx. 9 (Preface, 1)
  – limit 522-8
  – tabulated values, for cables Appx. 9
– exceeding low voltage, applicability of the Regulations to 11-2
– exceeding 250V, warning notices 514-4
– extra-low, definition p.12
– extra-low, description of range 11-2
– extra-low (*See also FELV, SELV*)
– for testing 613-5, 613-7
– high (*See High voltage*)
– limit–
  – ceiling roses 553-19
  – filament lampholders 553-15
  – HV discharge lighting 554-3, 554-6
  – safety separated circuits 413-36
– low–
  – definition p. 12
  – range 11-2
  – reduced 471-27 to 471-33
– marking, of different voltages within equipment 514-4
–nominal–
  – assessment of 313-1
  – definition p. 12
  – reduced low voltage circuits 471-28
  – SELV circuits 411-2, 411-10, 471-2, 471-3, 471-41
  – suitability of equipment for 512-1
– operated protective device (*See Fault-voltage operated...*)
– oscillation, in IT systems 413-13 Note
– ranges covered by the Regulations 11-2
– rating of fuses and circuit breakers 530-3
– surges, in discharge lighting circuits 521-1 Note 3
– tolerances (*See definition of Voltage, p. 12*)

W

Walls–
– fire-resistant, cables passing through 528-1
– insulating, resistance of 413-29
– outdoor, cables on 523-25
– socket outlets on, mounting height 553-8
– thermally insulating, cables in 522-6

Warning notices–
– caravans, in 514-6
– earthing and bonding connections 514-7, 542-18
– emergency switching 537-16
– HV discharge lighting 554-5, 554-17
– isolation of equipment 461-3
– voltages exceeding 250V 514-4
Warning signs, for areas reserved for skilled or instructed persons 471-25
Water–
– aggressive conditions of 554-27 Note
– heaters–
  – and boilers, electrode type 554-20 to 554-26
  – and standard final circuits Appx. 5
  – diversity Appx. 4 (Table 4B)
  – having immersed heating elements 554-27 to 554-30
  – 'instantaneous', in bathrooms 471-39
– immersion in, effect on body resistance Note to 471-2 & 471-3
– immersion in, effect on earth fault loop impedance 471-11 & Note
– pipes–
  – aluminium conductors with 547-1
  – bonding to, at points of contact 525-10
  – main bonding of 413-2, 547-1 to 547-3
  – public, prohibited as earth electrodes 542-14
  – segregation from 525-10, 525-11
– supply authority 554-30 Note 2
– wiring systems exposed to 523-7 to 523-15
Weather, equipment exposed to 13-17
Weather, wiring systems exposed to 523-6, Appx. 10
Weight (*See Mass*)
Wiring materials, selection and erection of Chap. 52
Wiring Regulations Committee, constitution p. v
Wiring systems–
– ambient temperature for 523-5
– corrosive or polluting substances, exposed to 523-17, 523-18
– damage by fauna 523-34
– livestock, inaccessibility to 523-34
– metalwork of, as protective conductor 543-9
– numbers of cables in 529-7
– selection of types Sec. 521, Appx. 10
– supports for Appx. 11
Workmanship 13-1